DIFFERENTIABLE DYNAMICS

DIFFERENTIABLE DYNAMICS

An Introduction to the Orbit Structure of Diffeomorphisms

Zbigniew Nitecki

THE M.I.T. PRESS
Cambridge, Massachusetts, and London, England

ISBN 0 262 14009 8 (hardcover)
ISBN 0 262 64011 2 (paperback)

Library of Congress catalog card number: 78-162818

PUBLISHER'S NOTE

The aim of this format is to close the time gap between the preparation of certain works and their publication in book form. A large number of significant though specialized manuscripts make the transition to formal publication either after a considerable delay or not at all. The time and expense of detailed text editing and composition in print may act to prevent publication or so to delay it that currency of content is affected.

The text of this book has been photographed directly from the author's typescript. It is edited to a satisfactory level of completeness and comprehensibility though not necessarily to the standard of consistency of minor editorial detail present in typeset books issued under our imprint.

The MIT Press

Manuscript received June 1971;
publication date October 1971.

CONTENTS

PREFACE

This book is a revised and somewhat expanded version of lecture notes written in 1969-70 and based on a course given by the author at Yale University in the fall semester, 1969. The notes, like the course, were meant to be an elaboration of the outline of the dynamical theory of diffeomorphisms given by Smale in his 1967 survey [3], with some updating of his picture of the "state of the field".

The presentation is reasonably self-contained, assuming an elementary acquaintance with Banach spaces and with differentiable manifolds, as well as a few facts about Lie groups in chapter 3. It should therefore theoretically be accessible to graduate students (in, say, topology) beyond their second year; in fact, the original course was taught primarily to second and third year graduate students at Yale. However, in practice, the "maturity" of the reader will prove important when it comes to the level of understanding the "meaning" of various theorems and techniques. Thus despite the considerable attention paid to the explanation and motivation of material in this book, the author feels compelled to rate it "X".

The material through Chapter 6 was the content of the 1969 graduate course at Yale, which was attended mainly by graduate students in their second and third years.

The exposition is organized around the development of the Ω -stability theorem, so that other basic phenomena in the dynamics of diffeomorphisms are viewed as motivation for the concepts involved in that theorem.

Other attempts have been made to unify the treatment here — for example, the use of Banach space techniques, especially the use of Hartman's theorem in chapter 3. However, the book still reflects its origin, in the ad hoc character of certain techniques of proof (e.g., chap. 4), the sometimes cavalier attitude toward specific estimates, and the total omission of details of certain important results (e.g., stability of Morse-Smale systems, the various counterexamples, the construction of Markov partitions). The original notes were intended as a relatively informal aid to workers in the field when teaching courses; however, it was felt that a somewhat detailed exposition of the important results in this area should be made available to a more general mathematical public in a form more accessible than scattered research papers. The author hopes that his picture of the reader has been a realistic one, and begs pardon for his many sins, both of commission and of omission.

A major area of sinning will probably be in the author's acknowledgement of credit for various concepts and results, especially with respect to the considerable work of east European mathematicians, primarily in the Soviet Union. In revising the original notes, a sincere attempt has been made to counteract the author's earlier overemphasis on the work of the mathematicians among whom he had "grown up"; however, it is likely that serious omissions remain. The author wants to apologize in advance for any such omissions, and to express his hope that future communication will eliminate such imbalance. The reader is referred to a forthcoming set of notes by Sinai [2] which have been announced (but which the author hasn't yet seen), and which will undoubtedly give an authoritative survey of Soviet and other east European work in the field.

Differentiable dynamics can be seen as the most recent fusion of topology with the "qualitative" theory of differential equations — two fields which have developed quite independently since Henri Poincaré first suggested many of the basic concepts of each in his study of celestial mechanics at the turn of the century. An older cousin of differentiable dynamics is "topological dynamics", another encounter of topology with the theory of flows; begun by Morse and Hedlund in the 1930's, this was an outgrowth of the work of G.D. Birkhoff, which in turn carried on Poincare's study of mechanics. Each field has its own interests: the topological dynamicists seem oriented toward "point-set" or more specifically metric-space topology, and so generally study homeomorphisms or continuous flows on metric spaces, while workers in differentiable dynamics often have their roots in differential topology, and thus study diffeomorphisms and smooth flows on manifolds. These differences influence the flavor of the two studies: while the study of homeomorphisms often involves the analysis of the topology of the underlying space and the very detailed analysis involved in "symbolic dynamics" techniques, differentiable dynamics has been suffused with ideas from differential topology: Whitney's topologizations of function spaces and investigation of dense sets of maps (e.g., transversality theory) are fundamental to the concept of genericity and the role played by the various mutations of Andronov-Pontrjagin's concept of structural stability; recently manifolds of mappings have become increasingly useful; and hyperbolicity conditions have been formulated in terms of vector bundles. The distinction is not, of course, absolute: the original work of Morse and Hedlund involved geodesic flows on manifolds of negative curvature, while the recent work of Sinai, Bowen, and Williams has made extensive use of symbolic techniques. Sinai, Anosov and Bowen (and, of course, many co-workers) have also related the theory of smooth systems to the other traditional viewpoint in analysis of dynamical systems, ergodic theory.

A broad and authoritative survey of the area of differentiable dynamics is Smale's 1967 paper [3]; a somewhat narrower vision of the field is sketched in this book. A brief characterization of the topic would mention the study of diffeomorphisms, smooth flows and to some extent smooth actions of groups on (usually) compact manifolds, regarded as dynamical systems - that is, involving the global analysis of invariant sets, recurrence and asympoticity properties - from the point of view of stability under perturbation and genericity in appropriate function spaces.

The specific material covered in this book is reflected in the titles of the various chapters and sections. Two introductory chapters deal, respectively, with the relations between discrete and continuous dynamical systems, and with basic function-analytic tools employed in the theory. Chapter 1 commences the dynamic theory with three studies of systems on the circle, where complications involving the topology of the underlying space are minimal. The first, "classical" section is an attempt to convey the flavor of a more traditional approach by restating Van Kampen's treatment of rotation numbers for homeomorphisms of S^1, including a proof of the Denjoy theorem. The second, "generic" section presents Peixoto's structural stability theorem for diffeomorphisms of S^1, while the third section treats Shub's "expanding maps" in a circle context. This treatment is continued at the end of the book, in Appendix I.

Chapter 2 contains the basic "generic" theory of periodic points: the Hartman linearization and stability theorems (both proven in a Banach space setting) and part of the Kupka-Smale genericity theorem, which is completed in Appendix II. Morse-Smale and gradient systems are defined, but results are stated without proof.

Chapters 3 and 4 present two important classes of examples: the Anosov diffeomorphisms and Smale's "horseshoe". Chapter 3 begins with a relatively detailed study of Thom's hyperbolic toral automorphisms (supplemented by Appendix III), proceeds with the Anosov stability theorem (proven as a corollary of the infinite-dimensional Hartman theorems) and closes with some further examples (infra-nilmanifold automorphisms) and conjectures. The openness of Anosov diffeomorphisms is proven in Appendix IV. Chapter 4 is a restatement of Smale's "horseshoe" paper [4].

Chapters 5 and 6 are devoted to the main steps in the Smale Ω-stability theorem. Chapter 5 gives general results on local stability and stable manifold theory for hyperbolic sets, following the basic paper of Hirsch and Pugh [1]. Chapter 6 contains proofs of the "local product structure" and "spectral decomposition" theorems, a discussion of Ω-explosions and the "filtration" construction, and the proof of the Ω-stability theorem itself. This chapter draws from the four papers among which the details of the full proof of Ω-stability are scattered.

Finally, chapter 7 is a quick survey of some results in differentiable dynamics established since the 1967 survey. Four major areas are covered: the various counter-examples (Smale, Abraham-Smale, Newhouse, Simon) to Smale's genericity conjectures; the stability theorems of Robbin, Sacker, and Hirsch-Pugh-Shub; the work of Sinai and Bowen on Markov partitions and invariant measures; and the work of Guckenheimer and Williams on the topological structure of basic sets.

I would like to mention a few of the idiosyncracies of this book, so they will at least not confuse the reader. I have adopted John Kelley's convention of numbering with a "zero" preliminary chapters or sections, which present material that is used in the body of the text but is not properly a part of it. By abuse of logic, I am generous in my non-rigorous use of quantifier symbols in place of phrases like "for all" ("$\forall$"), "there exist" ("$\exists$"), "there exists unique" ("$\exists !$") or "such that" ("$\ni$"). I also make liberal use of the contraction "iff" ("if and only if"). The word "id" or "id_X" refers to the identity of X - in the mapping, not Freudian sense. The real numbers, rationals, integers, and (strictly) positive integers are denoted, respectively, by $\mathbb{R}$, $\mathbb{Q}$, $\mathbb{Z}$, and $\mathbb{Z}^+$. Set-theoretic difference is denoted $A \sim B$. The end of a proof is indicated by the symbol //.

Many individuals have contributed extensively if indirectly to the writing of this book — whether by explaining material to me or discussing the treatment of certain topics, or in various supporting roles. I would like to thank S. Smale, M. Hirsch, C. Pugh, M. Shub, and P. Walters for mathematical help, and Mrs. Mary Ellen DelVecchio for the highly non-trivial contribution of 500-600 pages of nearly-perfect typing to the preparation of this volume. In the classic tradition, the responsibility for errors in the text is entirely mine.

Finally, I had intended to present students reading this book with a quote from the first verse and chorus of B. Dylan's "Ballad of a Thin Man", but the statement would probably be lost on the uninitiated, and them that knows, ain't saying — to paraphrase the little neighbor boy, who in turn was quoting Lao-Tzu.

Zbigniew Nitecki
April 1971
New Haven, Connecticut

DIFFERENTIABLE DYNAMICS

INTRODUCTION
FLOWS AND DIFFEOMORPHISMS

Historically, the theory of differential dynamical systems developed during the past decade by Smale, Peixoto, Hartman, and a long list of associates is an outgrowth of the qualitative theory of differential equations initiated by Poincaré and Birkhoff in the first part of this century. Loosely stated, the basic concern of these theories is to obtain a global "topological" picture of the behavior of solutions of differential equations on manifolds.

Such differential equations are given globally as sections of the tangent bundle, or vectorfields on the manifold. Locally, the tangent bundle of a manifold M modelled on a Euclidean (or Banach) space E is diffeomorphic to the trivial bundle

$$\begin{array}{c} U \times E \\ \downarrow \pi \\ U \end{array}$$

where U is an open subset of M (or of E), and π is the projection onto the first factor. A vectorfield X is given locally as a map $X: U \longrightarrow U \times E$ via

$$X(m) = (m, f(m)) \qquad (m \in U) .$$

The integral curves of the vector field are then given locally as solutions γ of the differential equation

$$\dot{\gamma}(m) = \frac{d\gamma}{dt}\Big|_m = f(m) .$$

Under certain general assumptions (in particular, if X is continuous and M compact), the uniqueness and existence theorems for ordinary differential equations let us "piece together" these local solutions to give a <u>flow</u> on M — that is, a map $\varphi : M \times R \longrightarrow M$ (denoted $\varphi(m,t) = \varphi_t(m)$) satisfying:

$$\text{(i)} \qquad \varphi_0 = \text{id}$$
$$\text{(ii)} \quad \varphi_t(\varphi_s(x)) = \varphi_{t+s}(x) .$$

However, while flows have been historically the main object of study in this area, the more recent workers have tended to concentrate on diffeomorphisms instead. To understand the relation between the two, we shall consider a number of ways of passing between flows and diffeomorphisms

1. TIME-ONE MAPS

The most obvious way that one can pass from a flow to a diffeomorphism is by considering the flow at some particular, discrete time t — that is, we consider the map φ_t for a fixed t. By the existence and uniqueness of solutions, the map φ_t is a diffeomorphism of M with itself (provided X is differentiable). The iterates of φ_t,

$$(\varphi_t)^n = \underbrace{\varphi_t \circ \cdots \circ \varphi_t}_{n \text{ times}} = \varphi_{nt} \qquad (n \in \mathbb{Z})$$

form an action of the group $\mathbb{Z}$ on M. This action amounts to "sampling" the action of the flow φ at discrete intervals of time, of duration $|t|$. Then the various dynamic properties of the flow φ (recurrent points, fixedpoints, asymptotic limits of orbits) are reflected in corresponding properties of the group of diffeomorphisms $\{\varphi_{nt}\}_{n \in \mathbb{Z}}$.

One might ask the converse question, known as the

Embedding Problem: Under what conditions can a given diffeomorphism $f: M \longrightarrow M$ be embedded as the time-one map of a flow $(f = \varphi_1)$?

Unfortunately, recent work of J. Palis indicates that most diffeomorphisms can't be embedded in flows.

Two observations make this result somewhat plausible. First, it is clear that φ_t is homotopic to φ_o = id, so for example orientation-reversing diffeomorphisms cannot be embedded. Second, on the circle it is easy to construct non-embeddable diffeomorphisms which are homotopic to the identity. One simply observes that a flow on the circle with a single closed orbit must be periodic (since that closed orbit is the circle), so that any diffeomorphism of S^1 with non-fixed periodic points which is not itself periodic (ie, $f^n(x_o) = x_o$ for some x_o, $n > 0$, but $f^n \neq$ id) cannot be embedded.

2. CROSS-SECTIONS AND SUSPENSIONS

Thus, under the embedding-time one relation, every flow gives rise to a diffeomorphism, but diffeomorphisms form a more general class, since many diffeomorphisms are not thus related to flows. A relation in which the opposite holds true arises from Poincaré's notion of cross-sections.

As a motivating example, consider the flow on the torus T^2 obtained by regarding T^2 as $\mathbb{R}^2/\mathbb{Z}^2$, with X the unit vectorfield in the x-direction. The flow, then, moves at constant speed along the "latitude" circles on the torus (Fig. 1).

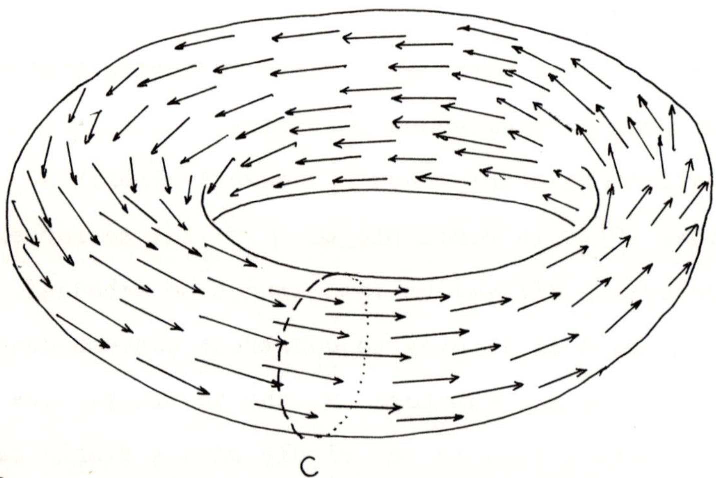

Figure 1

If we take C to be the projection on T^2 of the y-axis, we note that any point leaving the circle C returns to C after a positive time. We can define a map f on C to itself by letting $f(c)$ be the first point at which the orbit of $c \in C$ returns to C. For the particular vectorfield X above, this map is the identity; but if we let X be the vectorfield gotten from projecting the constant vectorfield $X_\theta = \sin\theta \dfrac{\partial}{\partial x} + \cos\theta \dfrac{\partial}{\partial y}$ $(\theta \neq \pm \frac{\pi}{2})$ on $\mathbb{R}^2$, then the corresponding map of C is just a "rotation" by $\tan\Theta$, and recurrence properties of X will be reflected in recurrence properties of $f: C \longrightarrow C$.

More generally, given a vectorfield X on a manifold M, a <u>cross-section</u> of X is a closed submanifold $\Sigma \subset M$ of codimension 1 (ie, of dimension one less than dim M) such that

(1) Σ is transverse to X
(ie, X and $t\Sigma$ generate $TM|\Sigma$)
(2) every orbit leaving Σ intersects Σ in both future and past time.
(3) every orbit intersects Σ.

Then if Σ is a cross-section of X, we can define the <u>first-return map</u> $f:\Sigma \longrightarrow \Sigma$ by setting

$f(x) = \varphi_{t_o(x)}(x)$,

where $t_o(x)$ is the least $t_o > 0$ for which $\varphi_{to}(x) \in \Sigma$. Note that this defines a diffeomorphism of Σ (by smooth dependence of solutions on parameters), and that condition (1) implies $t_o(x) > 0$.

Thus, a cross-section of a flow gives rise to a diffeomorphism on a submanifold. However, not all flows admit cross-sections. For example, condition (1) implies in particular that X has no singularities; by index arguments, this restricts even the topological type of M.

On the other hand, we have a nice converse, namely

Every diffeomorphism can be regarded as the first-return map on a cross-section of some flow.

This is accomplished by the mechanism of <u>suspension</u>.

Given $f: M \longrightarrow M$ a diffeomorphism, we construct a new manifold $\widetilde{M}$ as the quotient space of $M \times I$ (the product of M with the unit interval $[0,1]$) under the relation $(x, 1) \sim (f(x), 0)$.

The unit vectorfield on $M \times I$ pointing along I projects to a smooth vectorfield X on $\widetilde{M}$, and the 0-section $M_0 = M \times \{0\}$ projects to a cross-section of X with first return map f.

In the case $M = S^1$, this construction gives $\widetilde{M}$ either a torus or a Klein bottle, depending on whether f preserves or reverses orientation. In particular, a rotation through angle θ on S^1 gives the vectorfield X_θ discussed above as motivation for cross-sections.

We see, then, that the cross-section-suspension relation favors the diffeomorphisms, in that every diffeomorphism suspends in a flow, but not every flow has cross-sections. However the two kinds of relations we have considered indicate that the qualitative theory of flows and the dynamical study of diffeomorphisms should give rise to theories which, although differing in some details, would be of about equal generality and would develop along parallel lines.

In these notes, we shall henceforth be concerned almost exclusively with diffeomorphisms rather than flows, following the lead of Smale.

CHAPTER 0

PRELIMINARIES

0.1. FUNCTION SPACES

It will at times be useful to apply topological and even differential analysis to spaces of mappings from a manifold to itself. We collect here most of the concepts and results (without proof) that we might need for such an analysis. The most detailed reference for this section is the "Blue Notes" of Abraham [1]. Section 6 (pp. 776-81) of Eells' survey [2] does the same material in outline, while Eells [1] does a special case (C^0). For the material on spaces of sections, as well as generalizations, Abraham-Robbin [1] and Palais ([1], [2], [3]) have treatments.

We begin with a local topologization. Suppose X is an open subset of Banach space E, Y open in Banach space F.

Definition: Let $C^0(X,Y)$ = the collection of all continuous functions of X into Y.

We topologize $C^0(X,Y)$ be defining the uniform _ε-ball_ about $f \in C^0(X,Y)$ to be

$$\mathcal{B}^0_\varepsilon(f) = \{g:X \longrightarrow Y \text{continuous} \mid (\forall\, x \in X)\ \|f(x)-g(x)\| < \varepsilon\}.$$

The set of ε-balls ($\varepsilon > 0$) about functions $f:X \longrightarrow Y$ form a base for the uniform or C^0-topology. This is just the topology of uniform convergence on X.

To define the higher topologies, we review some calculus. Recall

<u>Definition:</u> $f: U \to F$ (where $U \subset E$ is open) is <u>differentiable</u> at $x \in U$ iff $\exists$ a bounded linear map <u>$Df_x: E \to F$</u> such that

$$\lim_{\|\delta\| \to 0} \frac{\|f(x+\delta) - f(x) - Df_x(\delta)\|}{\|\delta\|} = 0$$

Then, if f is differentiable at all points of U, we have a map

$Df : U \longrightarrow L(E,F)$

where $L(E, F)$ is the Banach space of (bounded) linear maps of E into F. If this map is continuous, we say f is <u>continuously differentiable</u>, or C^1 for short.

Note that Df is in turn a map from U to the Banach space $L(E,F)$; if this is differentiable, we get, for its derivative, the map

$D^2f: U \longrightarrow L(E, L(E, F))$.

But, in fact the space $L(E, L(E, F))$ can be identified with the space of (bounded) bilinear maps $E \times E \longrightarrow F$, and furthermore, the higher derivatives are actually symmetric as bilinear maps, so we really have a map

$D^2f: U \longrightarrow L_2^s\ (E, F)$

where $L_s^2(E, f)$ = (continuous) symmetric bilinear maps $E \times E \longrightarrow F$.

We can, of course, iterate this procedure to define higher derivatives

$D^r f: U \longrightarrow L_s^r(E, F)$

where $L_s^r(E, F)$ = (continuous) symmetric r-linear maps $E^r \longrightarrow F$. We say f is $\underline{C^r}$ in case $D^r f$ exists and is a continuous map.

We are now in a position to locally define the C^r topology. Suppose, as before, that $X \subset E$ and $Y \subset F$ are open subsets of Banach spaces.

<u>Definition:</u> $C^r(X,Y)$ = the set of C^r maps from X into Y. We topologize C^r by defining the <u>C^r-ε-ball</u> about $f \in C^r(X,Y)$ to be

$\mathcal{B}_\varepsilon^r(f) = \{g: X \longrightarrow Y \; C^r \mid (\forall\, k \leq r)(\forall\, x \in X) \; \| D_x^k - D_g^k \| < \varepsilon\}$.

Note:

(1) we adopt the convention

$D^0 f_x = f$

(2) this definition is equivalent to requiring that

$(\forall\, k \leq r)\; D^k g \in \mathcal{B}_\varepsilon^0 (D^k f)$

The C^r-ε-balls $(\varepsilon > 0)$ around C^r maps form a basis for a topology on $C^r(X,Y)$, called the <u>C^r topology</u>.

The definitions carry through as well for X a submanifold-with-boundary in E; if X is compact and Y a normed linear space, then letting $\|f\|_r$ be the sum of the suprema of norms of derivatives up to order r (ie, the sum of the terms on the right in defining $f \in \mathcal{B}_\varepsilon^r(0)$), we make $C^r(X,Y)$ into a normed linear space and if Y is complete, so is $C^r(X,Y)$ (this follows from the assertion that a uniform C^r-limit of C^r

functions is itself a C^r function). Thus, in this special case, the C^r topology is compatible with a Banach-space structure (cf. Eells, [2], section 6).

So far, we have defined topologies for spaces of mappings between subsets of Banach spaces; we proceed to globalize these local definitions. Suppose X and Y are now given manifolds, modelled (respectively) on Banach spaces E and F.

Given U, V coordinate patches on X, Y (resp.) and $f: X \longrightarrow Y$ a C^r map with $\overline{f(U)} \subset V$, we first let $\tilde{f}: U \longrightarrow V$ denote the local representative of $f|U$. Then we define

$\mathcal{B}^r_\varepsilon(U,V,f) =$ the set of C^r maps $g: X \longrightarrow Y$ satisfying:

(1) $\overline{g(U)} \subset V$;

(2) $\tilde{g} \in \mathcal{B}^r_\varepsilon(\tilde{f}) \subset C^r(U, V)$.

In other words, the <u>C^r-ε-ball</u> around f (with respect to coordinate patches U,V) is the set of all C^r maps sending U into V whose local representative has all derivatives up to order r within ε of the corresponding derivatives of $\tilde{f}$. As before, these balls (for all $\varepsilon > 0$, for all pairs of coordinate patches U,V, and all appropriate C^r maps f) form a basis for the <u>C^r topology</u> on $C^r(X,Y)$.

We remark in passing that $C^{r+1}(X,Y) \subset C^r(X,Y)$ has two topologies—the C^{r+1} topology, and the relative C^r topology. The C^r topology is the coarser of the two, so that C^r- open subsets of $C^{r+1}(X,Y)$ are also C^{r+1}- open. Furthermore,

$C^{r+1}(X,Y)$ is C^r- dense in $C^r(X,Y)$, so a C^{r+1}- dense subset of $C^{r+1}(X,Y)$ is also dense in $C^r(X,Y)$, and finally, the set $\mathrm{Diff}^r(M)$ is an open subset of $C^r(M,M)$ (this follows essentially from the fact that $GL(n, \mathbb{R})$ is open among matrices).

The constructions above allows us to apply topological techniques to these function spaces. However, at times the more delicate techniques of differential topology will prove useful; therefore we would like to endow $C^r(X,Y)$ with the structure of a differential (Banach) manifold. We assume henceforth that X is a compact and therefore finite-dimensional manifold.

The Banach space on which our manifold will be modelled is a space of sections of the tangent bundle, so we begin with this Banach structure. Let

$$\begin{array}{l} B \\ \downarrow \pi \\ X \end{array}$$

be a smooth vector bundle over X, with fiber E.

<u>Definition</u>: A C^r <u>section</u> of π is a C^r map $\gamma : X \longrightarrow B$ such that

$\pi \circ \gamma = id_X$.

We let $\Gamma^r(\pi)$ denote the space of C^r sections of π. Locally (ie, over a trivializing neighborhood U for the bundle) a section is given as a map $\gamma : U \longrightarrow U \times E$ via $\gamma(x) = (x, g(x))$. If the bundle has a Finsler structure (ie, a norm on fibers varying smoothly from fiber to fiber), then we can construct a metric on sections over U by defining $\|\gamma\| = \|g\|_r$, where the norm on the right is the C^r norm discussed above. (There may be problems in that U is not compact, but any section defined on all of X will surely have derivatives bounded on U.)

To get a global norm on $\Gamma^r(\pi)$, we pick a finite cover of X by trivializing neighborhoods, and take the norm of a section to be the supremum of the norms of its restrictions to these neighborhoods. One then shows that the norms obtained from two different covers of X by trivializing neighborhoods are equivalent, and furthermore, that any two Finsler structures on π are equivalent. (This heavily uses compactness!). So we have a

<u>Local Lemma</u>: The space $\Gamma^r(\pi)$ has a natural structure (up to linear homeomorphism) as a Banach space.

To extend this local Banach structure on $\Gamma^r(\pi)$ to a global manifold structure on the space $C^r(M,N)$, we need to draw on two concepts from differential topology. One is the exponential map of a Riemann metric: given a vector $v \in M_p$, we define $\exp_p(v)$ to be the point reached at time 1 by moving out along the geodesic tangent to v at p. On a small neighborhood of the origin in M_p, this is a diffeomorphism onto a neighborhood of p in M. When M is compact, the size of this neighborhood is uniform — that is, $\exists\ \varepsilon > 0$ independent of p such that the ε-ball at the origin of M_p is mapped diffeomorphically onto a neighborhood of p. We comment that the exponential map is as smooth as the tangent bundle.

Our second notion is that of the pullback of a tangent bundle: if $f: M \longrightarrow N$ is a smooth map and $TN \xrightarrow{\pi} N$ is the tangent bundle of N, then the <u>pullback</u> of TN by f, denoted <u>$f^* TN$</u>, is defined to be $\{(v,x) \in TN \times M \mid \pi v = fx\}$, with projection the map $\pi_*(v,x) = x$. We have a commutative diagram of maps

$$\begin{array}{ccc} f^* TN & \xrightarrow{\bar{f}} & TN \\ {\scriptstyle \pi_*}\downarrow & & \downarrow{\scriptstyle \pi} \\ M & \xrightarrow[f]{} & N \end{array}$$

where $\bar{f}$ is an isomorphism on fibers.

Now, to define the manifold structure of $C^r(M,N)$, we need to define the coordinate charts. By compactness arguments, there is a uniform $\delta > 0$ such that for any $p \in M$, the δ-ball about p is a diffeomorphic image of $\exp_p$; that is, if $\varepsilon > 0$ is the uniform radius such that $\exp_p | B_\varepsilon(0_p)$ is a diffeomorphism then $\exists\ \delta > 0$ such that $B_\delta(b) \subset \exp_p(B_\varepsilon(0))$. Then for any g C^0-δ-close to f, the map $\tilde{\varphi}_f(g): M \longrightarrow TN$ is defined by

$$\tilde{\varphi}_f(g)(x) = \exp_{f(x)}^{-1}(g(x)) .$$

Then using the fact that $\bar{f}$ is an isomorphism on fibers, we can pull back $\tilde{\varphi}_f(g)$ to a section $\varphi_f(g) \in \Gamma^r(f^* TN)$ by

$$\bar{f} \circ \varphi_f(g) = \tilde{\varphi}_f(g) .$$

It is fairly clear by standard arguments that, if f and g are C^r, then $\varphi_f(g)$ is a C^r- section of f^*TN, and that if g is C^r- close to f, then $\varphi_f(g)$ is close to the zero-section of f^*TN.

To get a slightly more intuitive picture of what this construction really involves, consider the graph of f. This can be regarded either as the point-set $\{(x, f(x)\}$ in $M \times N$, or as the function $gr(f) = (id, f) : M \longrightarrow M \times N$ given by $gr(f)(x) = (x, f(x))$. Intuitively, g is near f if the point-set $gr(g)$ is near the point-set $gr(f)$ — and in fact, a C^0-neighborhood of f is just that (see fig. 1).

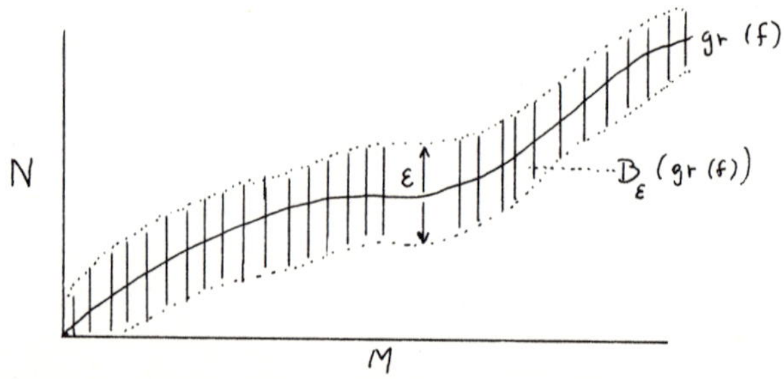

Figure 1. g $B^0_\varepsilon(f)$ iff $gr(g) \subset B_\varepsilon(gr(f))$

However, to refine our topology to the C^r sense, we need to go to mappings. The exponential map which we use in defining $\widetilde{\varphi}_f$ allows us to lift these graphs to $M \times TN$, where we can perform linear operations. Noting that $\widetilde{\varphi}_f(f)$ is just the 0-section ($\widetilde{\varphi}_f(f)(x) = 0_{f(x)}$), we see that, in effect, $\widetilde{\varphi}_f(g)$ is a linear representation of the "difference" $(f - g)$ — if that could be defined. And the idea behind saying "g is C^r-

close to f" is that this difference is C^r—small— ie, that the "difference" is close to 0. Finally, the use of $\bar{f}$ is simply a polite gesture, which translates this situation to that of sections of a vector bundle.

Now, to show that this procedure defines a manifold structure on $C^r(M,N)$, one needs to show that these local coordinates are in fact coordinates (ie, that $B^r_\varepsilon(f)$ maps homeomorphically onto an open set in $\Gamma^r(f^*TN)$), and that they "glue together" differentiably. We shall not go through the details here; the interested reader can check Abraham [1] or other references in Eells [2]. Essentially, the first of these steps involves recognizing that the C^r - topology on $\Gamma^r(f^*TN)$ and the C^r - topology on $C^r(M,N)$ make the map $\varphi_f(g)$ continuous, while the second step involves proving the differentiability of the map ω_f which we shall consider below. The exact result of these procedures is a

<u>Global Lemma</u>: For X a compact C^∞ manifold and Y a C^∞ Riemannian manifold, the space $C^r(X,Y)$ has a C^∞ differential structure modelled on a separable Banach space.

Given this manifold structure, we would like to develop a little calculus which might be useful in our applications. In particular, there are three operators on $C^r(M,N)$ whose derivatives would be useful to know. In the ensuing discussion, fix $f: W \longrightarrow X$ a C^r map of manifolds, compact whenever necessary to define the appropriate manifolds of maps.

First, consider the following

<u>Definition</u>: $\underline{\alpha_f}: C^r(X,Y) \longrightarrow C^r(W,Y)$ is defined by $\alpha_f(g) = g \circ f$.

We remark that this map is C^∞, and appeal for motivation to the observation that in the case of Y a Banach space (or open subset), α_f is a (bounded) linear map.

Next, consider

<u>Definition</u>: $\underline{\omega_f}: C^r(V,W) \longrightarrow C^r(V,X)$ defined by $\omega_f(g) = f \circ g$

We claim that this map is C^s whenever f is C^{r+s}. To motivate this statement (and help compute the derivative), we shall try to understand what the local representation of ω_f is.

Let $g \in C^r(V,W)$; then the chart φ_g at g is defined by commutativity of the following diagram

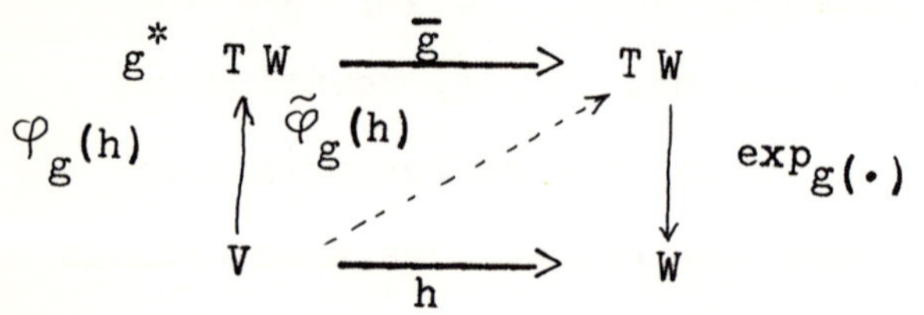

(1)

where h is near g in $C^r(V,W)$, and of course $\varphi_g(h)$ is a section of g^*TW.

On the other hand, local coordinates at $\omega_f(g) = f \circ g \in C^r(V,X)$ are given by the diagram

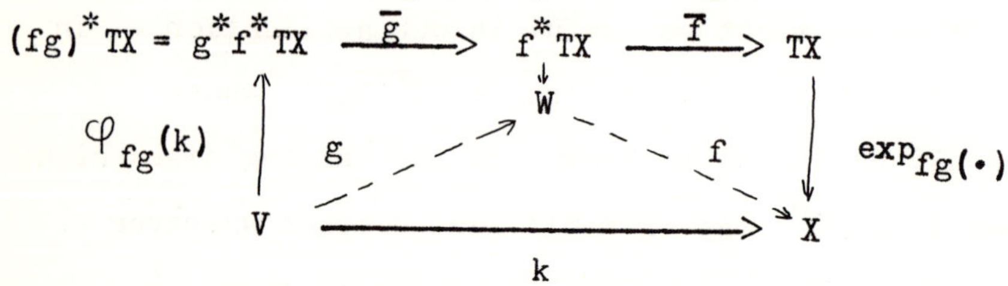

(2)

(where the dotted lines are not meant to commute with the rest).

Now, f naturally defines a map of diagram (1) into TX:

$$\begin{array}{ccccc} g^*TW & \xrightarrow{\ \bar{g}\ } & TW & \xrightarrow{\ \tilde{f}\ } & TX \\ {\scriptstyle \varphi_g(h)}\uparrow & & \downarrow{\scriptstyle \exp_g(\cdot)} & & \downarrow{\scriptstyle \exp_{fg}(\cdot)} \\ V & \xrightarrow[\ h\ ]{} & W & \xrightarrow[\ f\ ]{} & X \end{array} \qquad (3)$$

where $\tilde{f}$ is now defined to commute diagram (3). Then the local representation of ω_f, $\varphi_{fg} \circ \omega_f \circ \varphi_g^{-1} = \tilde{\omega}_f$, is given essentially by putting (2) and (3) together:

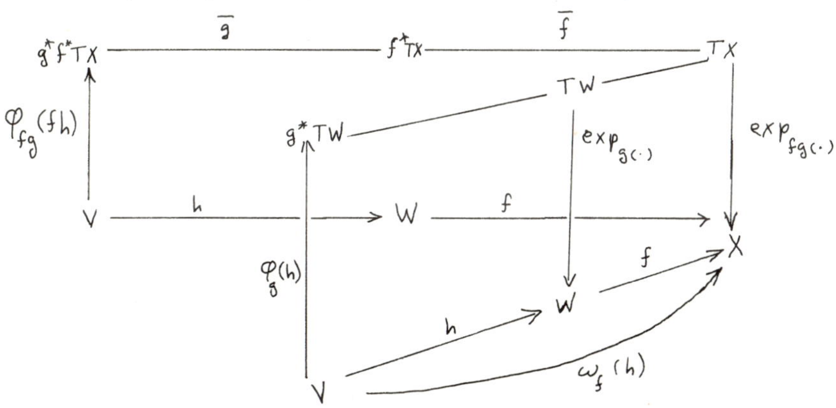

(4)

Written analytically, this says

$$\tilde{\omega}_f(\sigma) = \bar{g}^{-1} \circ \bar{f}^{-1} \circ \tilde{f} \circ \bar{g} \circ \sigma$$

for $\sigma \in \Gamma^r(g^*TW)$.

Now, when we differentiate $\tilde{\omega}_f$ with respect to σ, everything is essentially an identification except the map $\tilde{f}$, and its derivative is, almost by definition, the tangent map Tf of f. So, a linear approximation of $\tilde{\omega}_f$ is given by replacing diagram (3) with

$$\begin{array}{ccccc} g^{*}TW & \xrightarrow{\bar{g}} & TW & \xrightarrow{Tf} & TX \\ \downarrow{\scriptstyle g^{*}t} & & \downarrow{\scriptstyle t} & & \downarrow{\scriptstyle t} \\ V & \xrightarrow[g]{} & W & \xrightarrow[f]{} & X \end{array} \qquad (T3)$$

and this, plugged into (4), gives us the expression

$$\mathcal{D}\,\tilde{\omega}_f(0)(\sigma) = \bar{g}^{-1} \circ \bar{f}^{-1} \circ Tf \circ \bar{g} \circ \sigma \, .$$

This is essentially the expression for the derivative of ω_f at g. The above discussion can be more or less subsumed in the mnemonic

$$T\,\omega_f = \omega_{Tf}$$

— that is, the action of $T\,\omega_f$ on a section $\sigma \in \Gamma^r(g^{*}TW)$, regarded as an element of $T_g\, C^r(V,W)$, is (up to the appropriate identification) just post-composition with Tf. This result is proven more rigorously as the "Ω—lemma" in Abraham [1].

A third map which is useful to differentiate is the map

$$ev : C^r(X,Y) \times X \longrightarrow Y$$

given by $ev(f, x) = f(x)$. This again is differentiable (it is C^s if Y is a C^{r+s+2} manifold) — we remark that this map is (like α_f) locally linear in its first coordinate. We shall not need to worry explicitly about computing its derivative.

0.2. THE "GENERIC" VIEWPOINT

Having defined our topology, we can now proceed to outline Smale's approach to the orbit structure of diffeomorphisms.

19 The "Generic" Viewpoint

Classically, the basic problem of the qualitative theory of differential equations has been to classify the solutions of all differential equations in some "qualitative" (which usually means topological) way — to attach certain algebraic or topological invariants to a vectorfield which would determine "qualitative" features of the behavior of its flow. Examples of such theory are rotation numbers (see chapter 1) or the Poincaré-Hopf index for singularities of a vectorfield (and its relation to the Euler characteristic of the underlying manifold). However, these are generally very rough — for example, even the number of fixedpoints is not completely determined by the Poincare-Hopf index. It seems, furthermore, that the pathologies that can happen are so nasty that any idea of classifying all flows is utopian at present.

On the other hand, it might be possible to arrive at a classification if certain "exceptional" flows are excluded. Trying to understand the properties of flows as a class, one might get a first approximation by studying the properties of "most" flows. The naive definition of "most" here, given our topology, would be "a dense set". However, this notion is inadequate — for example, in such a definition, "most" numbers are rational — but "most" numbers are irrational as well! We are led to consider the notion of a "generic" property.

Definition: (1) A residual (or Baire) subset of a topological space is one which can be expressed as a countable intersection of dense, open sets.
(2) A property (of maps $M \longrightarrow N$) is said to be C^r-generic if it holds for all elements of a residual subset of $C^r(M,N)$. More specifically, a property of diffeomorphisms (vectorfields) on M is C^r-generic if it holds for a residual subset of $\mathrm{Diff}^r(M)$ ($\Gamma^r(TM)$).

Residual sets have the obvious property that a countable intersection of residual sets is still residual.

Definition: A topological space is called a Baire space if every residual subset is dense.

The usefulness of the notion of generic property becomes clear when we remark that a manifold modelled on a complete metric space is a Baire space. Our remark follows from the Baire category theorem (which states that complete metric spaces are Baire spaces) and the fact (an easy exercise) that a locally Baire space is globally Baire. In particular $\mathrm{Diff}^r(M)$, being an open subset of $C^r(M,M)$, is a Banach manifold and so is a Baire space.

Having carefully defined our notion of "most", we now want to clarify what constitutes a "classification". Two ideas come into play here. First, of course, we would like to define some concept of qualitative "equivalence" between two diffeomorphisms, from the point of view of orbit structure, and to determine equivalence classes of a residual set of maps. However, if we are to have a fighting chance of succeeding in such a classification scheme, we should be able to somehow take in large sets of diffeomorphisms with each equivalence class, so that the total number of classes is manageable. One might, for instance, hope that the equivalence classes would be open subsets of $\text{Diff}^r(M)$ — or at least open subsets of the residual set of diffeomorphisms which we are trying to classify.

The first notion of equivalence we try is, if we may wax categorical, the natural one of an "isomorphism" in the category of dynamical systems.

Definition: Two diffeomorphisms $f, g \in \text{Diff}(M)$ are topologically conjugate, denoted $f \sim g$, if there exists a homeomorphism $h: M \longrightarrow M$ such that the following diagram commutes:

$$\begin{array}{ccc} M & \xrightarrow{\ f\ } & M \\ h\downarrow & & \downarrow h \\ M & \xrightarrow[\ g\ ]{} & M \end{array}$$

(ie, we require that $g \circ h = h \circ f$, or $g = hfh^{-1}$). It is easily checked that topological conjugacy is an equivalence relation. We remark that the notion of differentiable conjugacy — that is, requiring h in the above definition to be a diffeomorphism — would lead to too stringent a concept of equivalence, for reasons that will become clear a little later.

To gain a picture of what topological conjugacy entails, we indicate some of its invariants:

1. $p \in M$ is a fixed point of $f \in \text{Diff}(M)$ if $f(p) = p$. We shall denote the set of fixedpoints of f by $\text{fix}(f)$. It is clear, then, that $\text{fix}(hfh^{-1}) = h(\text{fix}(f))$.
2. $p \in M$ is a periodic point of period $n \in \mathbb{Z}$ $(n \neq 0)$ if it is a fixedpoint of f^n. The set of all periodic points of f is denoted $\text{per}(f)$; the least period of $p \in \text{per}(f)$ is the least positive n such that $p \in \text{fix}(f^n)$. That

$\text{per}(h f h^{-1}) = h(\text{per}(f))$ follows from the observation that corresponding iterates of two topologically conjugate maps are conjugated by the same homeomorphism:

$$(h \cdot f \cdot h^{-1})^n = h \circ f^n \circ h^{-1}$$

3. The <u>orbit</u> of a point $p \in M$ under f is

$$O_f(p) = \{f^n(p) \mid n \in \mathbb{Z}\} .$$

The positive (negative) orbit of p is

$$O_f^+(p) = \{f^n(p) \mid n \in \mathbb{Z}^+\}$$

$$O_f^-(p) = \{f^{-n}(p) \mid n \in \mathbb{Z}^+\}$$

It is clear that

$$O^{(\pm)}_{hfh^{-1}}(p) = h[O_f^{(\pm)}(p)]$$

When dealing with a fixed f, we shall drop the subscript.
4. $p \in M$ <u>recurs</u> under f if p is in the closure of $O^+(p) \cup O^-(p)$. This means that either $p \in O^+(p) \cup O^-(p)$ (which means it is periodic) or p is a cluster point of $O(p)$. In the latter case, we say p is a <u>recurrent</u> point of f — equivalently, p is recurrent if

$$\exists n_i, \quad |n_i| \longrightarrow \infty \quad . \ni . \quad f^{n_i}(p) \longrightarrow p .$$

Clearly, recurrent points are taken into recurrent points by a conjugacy.
5. We say a point $p \in M$ <u>wanders</u> under f if it has a neighborhood with discrete orbit — that is, if $\exists U$, a neighborhood of p, such that

$$f^n U \cap U = \emptyset \qquad (\forall n \neq 0)$$

Then, we weaken the notion of recurrence by saying that $p \in M$ is <u>non-wandering</u> if for any neighborhood U of p,

$$(\exists n \neq 0) \qquad f^n U \cap U \neq \emptyset .$$

Equivalently, p is non-wandering if every neighborhood of p intersects some orbit twice. We call the set of all non-wandering points of f the <u>non-wandering</u> set of f, denoted $\Omega(f)$. It is closed, invariant set for f, and is a conjugacy invariant as well:

$\Omega(h \circ f \circ h^{-1}) = h[\Omega(f)]$.

(This because orbits of neighborhoods are in 1:1 correspondence under a conjugating homeomorphism).

In the direction of our second requirement for a classification scheme, we introduce the notion of stability, originally due to Andronov and Pontryagin [1].

Definition: $f \in \mathrm{Diff}^r(M)$ is structurally stable if there exists a neighborhood $\mathcal{U}$ of f in $\mathrm{Diff}^r(M)$ such that any $g \in \mathcal{U}$ is topologically conjugate to f.

We remark that structurally stable systems automatically form an open set in $\mathrm{Diff}^r(M)$.

We can now see why differentiable conjugacy would be too strong a condition: the corresponding notion of stability would be an empty one. For if $g = h \circ f \circ h^{-1}$ with $g, f, h \in \mathrm{Diff}(M)$, and if $p \in \mathrm{fix}\,(f)$, then $hp \in \mathrm{fix}(g)$ and $Tg_{hp} = Th_p \circ Tf_p \circ (Th_p)^{-1}$, where Tf_p is the linear transformation of M_p given by the derivative of f at p. This implies that the eigenvalues of Tg_{hp} and Tf_p are the same; but even a linear transformation can be perturbed so that its eigenvalues change — hence we can't hope to have differentiably stable maps.

We will often also bring into play a weaker notion of equivalence (and of stability) which however still preserves all of the invariants we mentioned above.

Definition: $f, g \in \mathrm{Diff}(M)$ are Ω-conjugate, denoted $f \underset{\Omega}{\sim} g$, if there is a conjugating homeomorphism between their non-wandering sets — ie, if there is a homeomorphism

$h: \Omega(f) \longrightarrow \Omega(g)$

making the following diagram commute:

$$\begin{array}{ccc} \Omega(f) & \xrightarrow{f|\Omega(f)} & \Omega(f) \\ h \downarrow & & \downarrow h \\ \Omega(g) & \xrightarrow[g|\Omega(g)]{} & \Omega(g) \end{array}$$

Note that $f \underset{\Omega}{\simeq} g$ means that the sets of fixedpoints, periodic points, and recurrent points for f are taken by h onto the corresponding sets for g.

The corresponding notion of stability is

Definition: $f \in \mathrm{Diff}^r(M)$ is (C^r) Ω-stable if $\exists\ \mathcal{U}$ a neighborhood of f in $\mathrm{Diff}^r(M)$ such that $g \in \mathcal{U} \Longrightarrow f \underset{\Omega}{\simeq} g$.

Given these definitions we might loosely define the "generic" approach of Smale by the following set of questions:

(1) Given a (compact) manifold M, find generic properties for diffeomorphisms of M.
(2) Given M, classify a residual subset of $\mathrm{Diff}^r(M)$, preferably of the set of topological (or Ω) - conjugacy classes of some residual set of diffeomorphisms of M with some more "computable" set — which is the same as giving a means of exhibiting a representative for each equivalence class. To make the problem practical, the elements of each equivalence class should probably be stable, in some appropriate sense.

We close this section with a few more "dynamical" definitions which will prove useful later on.

We fix $f \in \mathrm{Diff}^r(M)$, $x \in M$.

Definitions: (1) $\omega(x) = \{$cluster points of $O^+(x)\}$ (2) $\alpha(x) = \{$cluster points of $O^-(x)\}$; these are called, respectively, the ω - and α - limit sets of x; they are both closed and invariant.

Note: By "cluster" points here, we mean to include the

"degenerate" case of a periodic point — ie, $y \in \omega(x)$ if $\exists n_i \longrightarrow +\infty \ . \ni . \ f^{n_i}(x) \longrightarrow y$ (define $\alpha(x)$ <u>mutatis mutandis</u>)

<u>Definition:</u> A set $\mu \subset M$ is <u>minimal</u> (for f) if it is closed, invariant, and has no proper closed invariant subsets (other than $\emptyset$, of course).

Note that μ minimal, $x \in \mu$, then for each $x \in \mu$, either $\omega(x) = \emptyset$ or $\omega(x) = \mu$. In case our space is compact, $\omega(x) \neq \emptyset$ for any $x \in M$, so in that case, any point of a minimal set recurs.

Finally, we can also weaken the notion of conjugacy to

<u>Definition</u>: a <u>semi-conjugacy</u> of f to g is a continuous map h so that

$$\begin{array}{ccc} M & \xrightarrow{f} & M \\ h \downarrow & & \downarrow h \\ M & \xrightarrow[g]{} & M \end{array}$$

commutes; note that we do <u>not</u> require that h is a homeomorphism, or even onto. (In the latter case topological dynamicists call g a <u>factor</u> of f.)

Trivial, useless semi-conjugacies abound; if $g = id$, h a constant map, f can be anything; however, if h does have special properties, then it can tell us something about f, given information about g (and vice-versa).

0.3 A QUICK SURVEY OF TRANSVERSALITY THEORY

This material is taken, hastily, from Abraham-Robbin ([1]

chapter 4); a somewhat different formulation is given in Abraham, [1] and [2]. We state definitions and major results; proofs are in Abraham-Robbin.

Begin by noting that the graph of a map $f:X \longrightarrow Y$ can be regarded either as the point set $\{(x, f(x))\}$ in $X \times Y$ or as the natural homeomorphism of this set with X given by (id,f): $x \longrightarrow (x, f(x))$. By abuse of notation, we denote either the map or the set by $gr(f)$.

Recall also that if $A \subset B$ are linear spaces then A has <u>codimension</u> k if it has a k-dimensional complement — ie, if there is a subspace $C \subset B$ with $\dim C = k$ and $B = A \oplus C$.

We can then define the notion of transversality.

<u>Definition</u>: Let X,Y be C^1 (possibly infinite-dimensional) manifolds, W an embedded submanifold of Y, and $f:X \longrightarrow Y$ a C^1 map. Then we say f is <u>transverse</u> to W at $x \in X$ (denoted $f \pitchfork_x W$) iff either of the following two conditions holds:

(1) $f(x) \notin W$

(2) $y = f(x) \in W$, and

(a) $(T_x f)^{-1} [T_y W]$ has a closed complement (in the sense of linear spaces)

(b) $(T_x f)(T_x X)$ contains a closed complement to $T_y W$ in in $T_y Y$.

We say that f is transverse to W (denoted $f \pitchfork W$) if it is transverse to W at every point $x \in X$.

Sometimes, the notion of transversality is developed by starting with the transversality of two functions. In our terms, one says that $f:X \longrightarrow Y$ is transverse at $x \in X$ to $g:X \longrightarrow Y$ if $gr(f) \pitchfork_x gr(g)$, where we are really abusing our notation — $gr(f)$ is the function $(id,f):X \longrightarrow X \times Y$ and $gr(g)$ is the submanifold $\{(x, g(x))\}$ in $X \times Y$. This definition is equivalent to requiring that the image of Tf and the image of Tg span TY, the tangent bundle to Y, at least at points where the maps agree. One can then come back to our definition of $f \pitchfork W$ by asking that f be transverse to the embedding map $W \hookrightarrow Y$.

Some intuitive picture of the meaning of transversality can be gained from the following facts:

$f \pitchfork_x W$ iff, locally,

$f(x, x_2) = (\eta(x_1, x_2), x_2)$

ie, there are coordinate charts around x and $f(x)$ with the neighborhood of x given as $U_1 \times V_2$, the neighborhood of $f(x)$ given as $V_1 \times V_2$, with U_1, V_1 and V_2 open subsets of the appropriate subspaces, and with f projecting onto the identity in the second factor (see fig. 2).

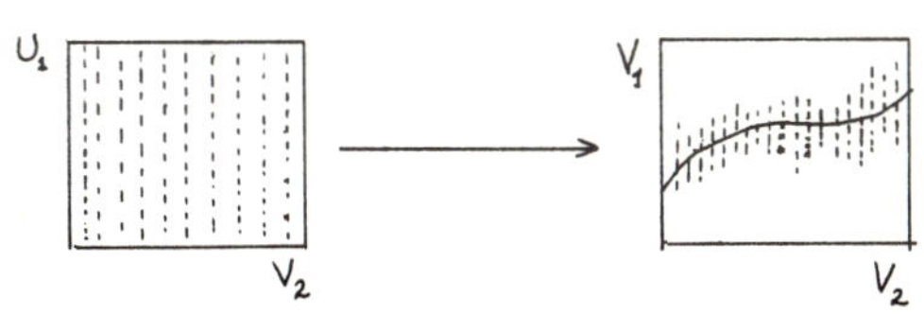

Figure 2. locally, $f(x_1,x_2) = (\ (x_1,x_2)x_2)$

The essence of the notion here is that by small perturbations in the V_2 - direction in the domain, one can move out in any direction complementary to V_1 in the range.

The notion of transversality allows us to reformulate neatly the

Implicit Function Theorem (for manifolds):

If $f:X \longrightarrow Y$, $W \hookrightarrow Y$ submanifold and $f \pitchfork W$, then:

(1) $\widetilde{W} = f^{-1}[W]$ is a submanifold of X
(2) codimension of $\widetilde{W}$ in X = codimension of W in Y
(3) $T\widetilde{W} = (Tf)^{-1}[TW]$.

The main reason for introducing transversality is its usefulness in finding generic properties of maps. This utility stems from the genericity, in very general circumstances, of the property of transversality itself. The prototype of such genericity theorems, due to Thom [1], says that for highly differentiable maps, transversality to a fixed submanifold is generic. We shall give here a more general formulation, due to Abraham. The point of the formulation is that we may be interested in transversality for only a limited class of maps which may (for example) form a nowhere dense collection among all C^r maps. So that we want to look at a map of some manifold into the space of all C^r maps, and ask whether those points whose corresponding map is transverse to some fixed submanifold form a residual set. We are led to the notion of a representation.

Definition: Suppose $\mathcal{A}$, X, Y are all C^r manifolds (possibly infinite-dimensional) with

$$\rho : \mathcal{A} \longrightarrow C^r(X,Y)$$

a map (so that to each $a \in \mathcal{A}$, ρ assigns a C^r map $\rho_a : X \longrightarrow Y$), then ρ is

(1) a <u>C^r-representation</u> if the "evaluation" map $ev_\rho : \mathcal{A} \times X \longrightarrow Y$ given by $ev_\rho(a,x) = \rho_a(x)$ is a C^r map (ie, the value of ρ_a depends C^r on both x and a).

(2) a <u>C^r-pseudo-representation</u> if for each $k \leq r$, the map

$$\rho^{(k)} : \mathcal{A} \longrightarrow C^{r-k}(T^kX,\ T^kY)$$

given by $\rho_a^{(k)} = T^k \rho_a$ is a C^0-representation (ie, if all of the derivatives, up to the $r^{\underline{th}}$, of ρ_a depend continuously on a — thus, we do not require that $\rho_a(x)$ depends differentiably on a).

<u>Examples</u>:

(1) suppose $\begin{array}{c} E \\ \downarrow \pi \\ X \end{array}$ is a C^r bundle with X a compact manifold. The space of C^r sections of π is a subspace of $C^r(X,E)$; let

$$\mathcal{A} = \Gamma^r(\pi), \quad Y = E, \quad \text{and}$$
$$\rho : \Gamma^r(\pi) \hookrightarrow C^r(X,Y)$$

be the set-theoretic inclusion. Then ρ is a C^r representation and clearly, any C^r-representation is a C^r-pseudo-representation.

(2) On the other hand, any continuous semi-group of C^r diffeomorphisms of a manifold is a C^r pseudo-representation, but only a C^0 representation. For example, taking

$\rho : \mathbb{R} \longrightarrow C^{\infty}(\mathbb{R}, \mathbb{R})$ via $\rho_a(x) = x + |a|$,

we see easily that $ev_\rho : \mathbb{R} \times \mathbb{R} \longrightarrow \mathbb{R}$ is not differentiable with respect to a, although

$\rho_a^{(k)} = (\rho_a, \mathrm{id}, \mathrm{id}, \ldots)$.

Abraham's results on transversality are then contained in the following theorems:

Theorem 1: (Openness of Transversality):

(a) (Openness of Non-Intersection):
$\mathcal{A}$, X, Y topological manifolds, $W \subset Y$ a closed submanifold, $K \subset X$ compact, if $\rho : \mathcal{A} \longrightarrow C^0(X,Y)$ is a C^0 representation, then $\mathcal{A}^0_{KW} = \{a \in \mathcal{A} \mid \rho_a[K] \cap W = \emptyset\}$ is open in $\mathcal{A}$.

(b) (Openness of Transversal Intersection)
$\mathcal{A}$, X, Y C^1 manifolds, X finite-dimensional, $W \subset Y$ a closed C^1-submanifold, $K \subset X$ compact,
$\rho : \mathcal{A} \longrightarrow C^1(X,Y)$ a C^1 pseudo-representation
then $\mathcal{A}_{KW} = \{a \in \mathcal{A} \mid (\forall x \in K)\ \rho_a \pitchfork_x W\}$ is open in $\mathcal{A}$.

(c) Furthermore, if ρ is actually a C^1 representation, X need not be assumed finite-dimensional.

Theorem 2: (Density of Transversality):

$\mathcal{A}$, X, Y, C^r manifolds

$\rho : \mathcal{A} \longrightarrow C^r(X,Y)$ a C^r-representation

$W \subset Y$ a submanifold,

$ev_\rho : \mathcal{A} \times X \longrightarrow Y$ the evaluation map.

Suppose

(1) dim X = n, codim W = q
(2) $\mathcal{A}$ & X are second-countable
(3) $r > \max(0, n-q)$
(4) $ev_\rho \pitchfork W$

then

$\mathcal{A}_W = \{a \in \mathcal{A} \mid \rho_a \pitchfork W\}$ is residual in $\mathcal{A}$.

Both proofs are given in Abraham-Robbin [1]. We remark that the bound (3) for r is that necessary for Sard's theorem, which is used in proving density.

CHAPTER 1
THE CIRCLE

The simplest non-trivial example of a compact manifold is the circle. It thus provides a setting for introducing the theory of dynamical systems without the topological complications which arise in higher dimensions. Even in this simple setting, the basic problems and techniques of dynamics come into play, yielding a theory of some interest in itself. On the other hand, the completeness of the results obtainable on S^1 is deceptive, as recent counterexamples to their generalizations in higher dimensions show.

We shall consider three approaches to dynamical systems on the circle. We begin with the classical theory of homeomorphisms of S^1, initiated by Poincaré and extended greatly by Denjoy; we then present Peixoto's work in [1], interpreted as a "generic" theory of diffeomorphisms of S^1; finally, we extend our study beyond diffeomorphisms to consider Shub's expanding maps on S^1.

1.1. THE CLASSICAL THEORY

Many aspects of this theory were understood by Poincaré, but the effect of differentiability assumptions was investigated by Denjoy ([1] - [4]); a full exposition of the theory is given by Van Kampen ([1]) and largely repeated in Coddington-Levinson ([1]), although the proof of Denjoy's theorem presented in both references contains a slight mistake. Finally, a different proof of Denjoy's theorem is given by Siegel ([1]) and repeated in Nemytskii-Stepanov ([1]), pp. 62-70).

Most of our analysis will be carried out under the assumption that all homeomorphisms we look at preserve orientation. At the end of the section, we shall return to this assumption and see what we can say about the orientation-reversing case. If $f: S^1 \longrightarrow S^1$ is an orientation-preserving homeomorphism of the circle, it can be "lifted" to a map $F: \mathbb{R} \longrightarrow \mathbb{R}$ which covers f via the covering projection $\exp: t \longrightarrow e^{2\pi it}$. Then F will satisfy:

(1) F is monotone increasing
(2) $F - id$ is periodic, with period 1.

We will refer to any such F as a <u>lift</u> of f; it is easy to see that two lifts of the same homeomorphism of S^1 will differ by an integer translation.

We wish to attach to each homeomorphism a number which in some sense measures the average "rotation" of points under f. We begin with a given lift.

<u>Lemma</u>: Suppose $F: \mathbb{R} \longrightarrow \mathbb{R}$ satisfies conditions (1) and (2) above. Then

$$\lim_{n \to \infty} \frac{F^n(t)}{n}$$

exists for each $t \in \mathbb{R}$, and is independent of t.

(Note: here, as before, $F^n = \underbrace{F \circ F \circ \dots \circ F}_{(n \text{ times})}$)

Proof

(1) Independence of t:

By the above remarks, $F - id$ is periodic, so

$|F^n x - F^n y| \leq |(F^n x - x) - (F^n y - y)| + |x-y| \leq |x-y| + 1$

so that $\lim\limits_{|n| \to \infty} \left(\frac{F^n x}{n} - \frac{F^n y}{n}\right) = 0$; hence the limit, if it exists, of $\frac{F^n(t)}{n}$ is independent of t.

(2) Existence if f has a periodic point:

A periodic point of period m is "covered" by a point $y \in \mathbb{R}$ with the property that $F^m(y) = y + p$. But then $F^{mn}(y) = y + np$ so that

$$\lim_{|m| \to \infty} \frac{F^{mn}(y)}{mn} = \lim_{|m| \to \infty} \frac{y + np}{mn} = \frac{p}{m} .$$

Now, since every k can be written

$k = rm + q \quad , \quad q < m$;

we note that $\frac{|F^k(y) - F^{rm}(y)|}{k} \leq \frac{M+1}{k}$, where

$M = \max\limits_{q < m} |F^q(y) - y|$ is a fixed number, so that

$$\lim_{|k| \to \infty} \frac{F^k(y)}{k} = \lim_{|n| \to \infty} \frac{F^{mn}(y)}{mn} = \frac{p}{m}$$

Hence, if f has periodic points, this limit exists, and is rational.

(3) Existence if f has no periodic points:

If there are no periodic points for f, then $F^m(x) - x$ is not an integer, for any $m > 0$ and any $x \in \mathbb{R}$. Hence, for every $x \in \mathbb{R}$ and every integer $m \neq 0$ there exists an integer p_m such that

$$p_m + x < F^m(x) < p_m + x + 1$$

or

$$p_m < F^m(x) - x < p_m + 1$$

and in fact p_m depends only on m, (not on x) by continuity considerations.

Adding together the above inequalities for $x = 0, F^m(0), \ldots, F^{m(n-1)}(0)$, we get

$$n\, p_m < F^{mn}(0) < n(p_m+1) \tag{1}$$

But the original inequality, for $x = 0$, reads

$$p_m < F^m(0) < p_m + 1 \tag{2}$$

Dividing (1) by mn, subtracting (2) divided by n, yields

$$\left| \frac{F^{mn}(0)}{mn} - \frac{F^m(0)}{m} \right| < \frac{1}{|n|} \tag{3}$$

Interchanging the roles of m and n, we get

$$\left| \frac{F^n(0)}{n} - \frac{F^m(0)}{m} \right| < \frac{1}{|m|} + \frac{1}{|n|}$$

Hence, the sequence $\frac{F^k(0)}{k}$ is Cauchy, and so has a limit. //

The proof of the lemma is thus complete; we shall refer to this limit as the <u>rotation number</u> of the lift,

$$\tau(F) = \lim_{|n| \to \infty} \frac{F^n(x)}{n} .$$

The relevance of this number to our study of conjugacy invariants is asserted in

Proposition 1: Let $\tau(F) + \mathbb{Z}$ denote the residue class of $\tau(F)$, modulo the integers. Then

(1) $\tau(F) + \mathbb{Z}$ is independent of the lift: if F, $\tilde{F}$ are two lifts of $f: S^1 \longrightarrow S^1$, then

$\tau(F) \equiv \tau(F) \pmod{\mathbb{Z}}$.

(2) $\tau(F) + \mathbb{Z}$ is invariant under order-preserving conjugacy: if

$g = h\, f\, h^{-1}$

h order-preserving, then

$\tau(f) = \tau(g)$.

Proof

(1) Any two lifts of f differ by a translation — ie, if F and $\tilde{F}$ lift the same map of S^1, then $\tilde{F}(x) = F(x) + k$.

Hence $\lim_{|n| \to \infty} \frac{\tilde{F}^n(x)}{n} = \lim_{|n| \to \infty} \frac{F^n(x) + nk}{n} = \lim_{|n| \to \infty} \frac{F^n(x)}{n} + k,$

and $\tau(F) + \mathbb{Z}$ depends only on the homeomorphism f of S^1 being lifted — so we can adopt the notation

$\tau(f) = \tau(F) + \mathbb{Z}$,

for F any lift of f.

(2) Now, given F lifting f and h an order-preserving conjugacy ($g = h\, f\, h^{-1}$), we want to find a convenient lift $\hat{F}$ of $h\, f\, h^{-1}$. Since h itself is an order-preserving homeomorphism, it can be lifted to a map H of the real line (monotone increasing, H -id periodic). We can pick the lifts F and H so that they take 0 into the closed unit interval.

It is clear that, taking

$\hat{F} = H\ F\ H^{-1}$

we have $\hat{F}$ a lift of g, and

$$\lim_{|n| \to \infty} \frac{\hat{F}^n(0)}{n} = \lim_{|n| \to \infty} \frac{H\ F^n H^{-1}(0)}{n} .$$

Since H takes 0 into $[0, 1]$ and H -id has period 1, we conclude

$$|\hat{F}^n(0) - F^n H^{-1}(0)| < 2$$

independently of n. Hence

$$\tau(\hat{F}) = \lim_{|n| \to \infty} \frac{H\ F^n H^{-1}(0)}{n} .$$

$$= \lim_{|n| \to \infty} \frac{F^n H^{-1}(0)}{n}$$

and by independence of the rotation number from the point chosen to compute it,

$$\tau(F) = \lim_{|n| \to \infty} \frac{F^n(H^{-1}(0))}{n}$$

$$= \tau(\hat{F}) .$$

This (together with part (1)) completes the proof of the proposition. //

We now adopt the notation $\tau(f)$, the "rotation number" of a homeomorphism f of S^1, for the residue class $\tau(F) + \mathbb{Z}$, F any lift of f.

Given this invariant of (ordered) conjugacy classes, we proceed to investigate its usefulness in telling us about the dynamic properties of the map. Note that in the process of defining the rotation number, we showed that the presence of a periodic point forced τ to be rational. We now show the converse, namely

Proposition 2: $\tau(f)$ is irrational if and only if f has no periodic points.

Proof

It suffices to show the "if" part, by our remarks above. Begin by noting that for any lift F of f,

$$\tau(F^m + k) = m\,\tau(F) + k$$

where $(F^m + k)(x) = F^m(x) + k$. Hence, it will suffice to show that $\tau(F) = 0 \Longrightarrow f$ has a fixed point, since if $\tau(F)$ is rational, there is an "m" and a "k" so that $\tau(F^m+k) = 0$.

Supposing the contrary ($\tau(F) = 0$ with no fixed points), we can assume, say, that $F(x) > x$ for every $x \in \mathbb{R}$. Then either $F^n(0) < 1$ for every n, or $F^k(0) > 1$ for some k. But in the latter case $F^{2k}(0) > F^k(1) > 2$, etc., so that

$$\frac{F^{nk}(0)}{nk} > \frac{1}{k}$$

and hence $\tau(F) > \frac{1}{k}$; since $\tau(F)$ was assumed to be zero, the first case must hold; thus, $\{F^n(0)\}$ is a monotone, increasing sequence bounded by 1. It therefore converges to some limit, which must project to a fixed point of f. It follows that $\tau(F) \in \mathbb{Z} \Longrightarrow f$ has periodic points, and the proposition is proven. //

Thus, rationality of the rotation number is equivalent to the existence of periodic points. We proceed to study the case of irrational rotation number — that is, the case when the system is free of periodic points.

Recall that for $x \in S^1$, $\omega(x)$ was defined to be the set of accumulation points of the sequence $\{f^n x\}_{n \in \mathbb{Z}^+}$.

Proposition 3: Suppose $\tau(f)$ is irrational. Then:
(a) $\omega(x)$ is independent of x
(b) If E denotes the closed invariant set $E = \omega(x)$ (any $x \in S^1$), then E is either perfect and nowhere dense, or else is the whole circle.

Proof

To prove part (a), we begin with a

Claim: given $x \in S^1$ and m, n distinct integers, let $I = [f^m(x), f^n(x)]$; then any forward orbit intersects I:
$(\forall y \in S^1)(\exists k > 0) \quad f^k y \in I$.

Proof of claim:

The intervals $f^{-k(m-n)}I$ abut (touch only at endpoints), so either they fill up S^1 after a finite number of iterates, or their endpoints converge monotonically to a fixed point of f^{m-n}. The latter possibility, however would contradict irrationality of τ, so y is an element of some $f^{-k(m-n)}I$, or $f^{k(m-n)}y \in I$.

Proof of (a):

Now to see that $\omega(x)$ is independent of x, note that if we have a sequence $\{f^{a_n}(x)\}$ of iterates of x converging to $x_0 \in \omega(x)$, and $a_n \longrightarrow +\infty$, then we can use the claim to pick $b_n \longrightarrow +\infty$ with $f^{b_n}(y) \in [f^{a_{n-1}}x, f^{a_n}x]$, and so

$f^{b_n}y \longrightarrow x_o$ and $x_o \in \omega(y)$. Thus $\omega(x) \subset \omega(y)$, and reversing the roles of x and y in the argument, $\omega(x) = \omega(y)$.

Proof of (b):

To see that E is perfect, we note that if $z \in E$, then $z \in \omega(z)$, so $z = \lim f^{a_n}z$, which are all distinct since z cannot be periodic.

Also, it is clear that by definition E is the unique minimal set. On the other hand, since f is a homeomorphism, the topological boundary of E is a closed invariant subset of E; thus, it is either empty $(E = S^1)$ or else it equals E, in which case E is nowhere dense. //

We distinguish the two possible cases of irrational rotation number by calling f <u>transitive</u> if $E = S^1$, and <u>intransitive</u> if not. Our next step is to investigate the dynamical significance of the rotation number itself (as opposed to just the property of rationality). Let ϑ_τ denote rotation through τ , defined as the projection to S^1 of translation by τ in $\mathbb{R}$. It is clear that ϑ_τ is uniquely defined by $\tau + \mathbb{Z}$, and that its rotation "number" is $\tau + \mathbb{Z}$. A partial converse is given by

<u>Proposition</u> 4: Let $\tau = \tau(f)$. Then $f| \Omega(f)$ is topologically semi-conjugate to an invariant subset of ϑ_τ. The semi-conjugacy is at most 2:1 on $\Omega(f)$, preserves orientation, and if τ is irrational, extends to a semi-conjugacy on all of S^1 which collapses the closure of intervals in the complement of $\Omega(f)$.

Corollary: If f is transitive, then f is topologically conjugate to ϑ_τ.

Proof

We separate the rational and irrational cases.

(1) Suppose $\tau(f) = \frac{M}{N}$, written in lowest terms. Observe that $\Omega(f)$ is just the set of periodic points and that all periodic points have least period N. Pick $p_0 \in per(f)$, and let $p_n = f^n(p_0)$. These points lie in a certain order around the circle; let p_i = successor of p_0 in this order. We

Claim:

(1) $\tau(f^i) = \frac{1}{N}$

(2) $f^{Mi} \mid \Omega(f) = f \mid \Omega(f)$.

Proof of claim:

For the first statement, note that the successor of p_i must be p_{2i}, and so on, so that the iterates of p_0 under f^i are successors. Thus, we can find a lift F of f^i such that $F^N(0) = 1$ (with 0 covering p_0). Hence, $\tau(F) = \lim_{k \to \infty} \frac{F^{Nk}(0)}{Nk} = \lim_{k \to \infty} \frac{k}{Nk} = \frac{1}{N}$. For the second statement, note that f^i preserves order, so that if q is any other periodic point and lies between p_0 and p_i, then $f^{ki}(q)$ lies between p_{ki} and $p_{(k+1)i}$. In particular, $f^{Mi}(q)$ is the unique iterate of q lying between p_1 and p_{i+1}, so must equal $f(q)$.

Now, to construct the conjugacy we note by (1) that it suffices to construct it assuming $M = 1$ (ie, in general, to conjugate f^i with $\vartheta_{1/N}$.) But this is easy: the intervals

$[p_{ki}, p_{(k+1)i}]$ abut, so taking a homeomorphism h of $[p_o, p_i]$ with $[1, \exp \frac{1}{N}]$, we see that the restriction of h to $[p_o, p_i] \cap \mathrm{per}(f)$ extends to a conjugacy of $\mathrm{per}(f)$ with an invariant subset for $\vartheta_{1/N}$.

(2) Suppose τ is irrational. We begin with a

claim: τ irrational, $x \in \mathbb{R}$, F a lift of f, then

$n_1 \tau + m_1 < n_2 \tau + m_2 \quad (m_i, n_i \in \mathbb{Z})$

iff $F^{n_1}(x) + m_1 < F^{n_2}(x) + m_2$

ie, the map $n\tau + m \longrightarrow F^n(x) + m$ is an order-presrrving injection.

Proof of claim:

We note that the order of the $F^n(x) + m$ is independent of x, since $F^n(x) - F^k(x) \notin \mathbb{Z}$ ($\forall x$). Hence, it suffices to prove the lemma for $x = 0$. But in this case

$F^n(0) + m < F^k(0) + \ell \iff$

$F^{n-k}(0) < \ell - m \iff$

$\tau(n-k) < \ell - m$.

(The last "iff" holds because

$F^{n-k}(0) - 0 < \ell - m \Longrightarrow F^{p(n-k)}(0) - 0 < p(\ell - m)$

so that $\tau = \lim_{p \to \infty} \frac{F^{p(n-k)}(0)}{p(n-k)} < \frac{\ell - m}{n-k}$.)

Now, to construct the semi-conjugacy, we pick $p_o \in E = \Omega(f)$, and let y_o cover it in $\mathbb{R}$. We pick F a lift of f, and let $y_n = F^n(y_o)$. Then taking $B = \{y_n + m\}$, $A = \{n\tau + m\}$, $(m, n \in \mathbb{Z})$, the map $g\colon B \longrightarrow A$ via $g(y_n + m) = n\tau + m$ extends to a unique order-preserving (in fact, continuous) map $g\colon \mathbb{R} \longrightarrow \mathbb{R}$ with the properties:

(1) onto S^1
(2) $g^{-1}A = B$.

In addition, g satisfies

(1) $g(x+1) = g(x) + 1$
(2) $gF(x) = g(x) + \tau$.

Hence, g clearly projects to a semiconjugacy. We remark that the property $g^{-1}A = B$ implies in particular that g fails to be injective only on an interval $I = [e_1, e_2]$ with the property that e_1 is a limit from below and e_2 a limit from above of $O^+(x_0)$, and $\mathring{I} \cap O^+(x)$ is empty. In this latter case g collapses I to a point, but $I \cap \Omega(f) = \partial I$. Hence, f is at most 2:1 on $\Omega(f)$, and in the transitive case, since no such interval exists, g is a continuous injection, hence a homeomorphism. Finally, the fact that A is dense in $\mathbb{R}$ (since τ is irrational) implies that g is onto S^1 in the irrational case. //

This proposition gives us considerable insight into the relation of the rotation number to the orbit structure of a homeomorphism.

We can, in fact, characterize the rotation number as the unique angle whose rotation contains a factor of $f|\,\Omega(f)$. Alternately, if we regard the circle as the multiplicative group $\{e^{it}\}$, then for any lift F of f, multiplication by $e^{2\pi i\,\tau(F)}$ is the unique rotation containing a factor of $f|\,\Omega(f)$.

We can use this semi-conjugacy to build fairly accurate models of the dynamic behavior of various cases. In the case of rational rotation number, the non-wandering set is relatively unrestricted: given any closed subset of the arc $[0, \exp \tau)$, we can (by including its translates into each interval $[\exp n\tau, \exp(n-1)\tau))$ embed it as part of $\Omega(f)$, for f a homeomorphism with rotation number τ. On the other hand, since the semi-conjugacy in the rational case is a homeomorphism, we can characterize the map on the rest of the circle: $S^1 \sim \Omega(f)$ consists of open intervals; each is mapped onto itself under f^M, in fixed point-free manner. But an interval without fixed points must have either $f^m(x) > x$ or $f^m(x) < x$, uniformly over the interval. Thus, on any given interval in $S^1 \sim \Omega$, every point is shifted somewhat clockwise (or counter clockwise) by f^M.

For irrational τ, there are two cases, transitive and intransitive. In the transitive case, f is just (up to change of coordinates) a rotation by τ. In the intransitive case, each interval of $S^1 \sim \Omega$ has a disjoint orbit, and in fact f can be imagined as obtained from the rotation ϑ_τ by "splicing in" an interval for each point of some invariant set, and taking some homeomorphism on each of these intervals. This is the basis of Denjoy's example, below.

So far, this theory applies to all (orientation-preserving) homeomorphisms of the circle. In a series of papers in the

1930's and 40's, Denjoy ([1], [2], [3]) investigated the effect of differentiability assumptions on this theory, particularly on the topological type of $\Omega(f)$.

In the case of rational rotation number, differentiability conditions impose no restriction: by taking for our homeomorphism on intervals of $S^1 \sim \Omega$ a diffeomorphism C^r-near the identity, we can repeat the above construction to obtain any closed, invariant subset of ϑ_τ as the non-wandering set of a C^r diffeomorphism.

The more interesting case, however, is that of irrational rotation number. Denjoy shows by means of an example (which we defer to the end of the section) that a continuously differentiable homeomorphism can be either transitive or intransitive. However, with slightly stronger conditions, he shows that the intransitive case is the "pathological" one:

Theorem (Denjoy): If f is a diffeomorphism of S^1, with f' continuous and of bounded variation, and with $\tau(f)$ irrational, then f is transitive.

The proof we give is essentially that of Van Kampen in [1], modified somewhat to avoid a difficulty in his presentation.

Proof:

Begin by noting that for any lift F of f, our hypothesis implies that F' exists, is continuous, and is of bounded variation on $[0, 1]$.

We now wish to have some notion of the ordering of the iterates of any given point around the circle. We

<u>Claim</u>: Let $p_o \in S^1$; denote $f^i(p)$ by p_i. Suppose n has the property that for $|k| < |n|$; p_k does not lie between p_o and p_n. Then the two finite sequences

$$\left.\begin{array}{l} p_o, p_1, \ldots, p_{n-1} \\ p_{-n}, p_{1-n}, \ldots, p_{-1} \end{array}\right\}$$

alternate around the circle (in the sense that no point of either sequence is interior to the arc (p_k, p_{k-n})).

We use the semi-conjugacy of proposition 4 to establish the claim: the property is clearly true of ϑ_τ, and since the semi-conjugacy preserves order, it is also true of f.

Consider now the function $\varphi(x) = \log F'(x)$. This is continuous, of period 1, and of bounded variation in $[0, 1]$. If $x_i \in \mathbb{R}$ covers $p_i \in S^1$, then we can infer the existence of a bound, V, such that for any n satisfying the hypothesis of the claim,

$$|[\varphi(x_o) + \varphi(x_1) + \ldots + \varphi(x_{n-1})]$$
$$- [\varphi(x_{-n}) + \ldots + \varphi(x_{-1})]| < V$$

or in other words,

$$|\log (\frac{d}{dx} F^n(x_o) \cdot \frac{d}{dx} F^{-n}(x_o)| < V$$

or

$$e^{-V} < \frac{d}{dx} F^n(x_o) \cdot \frac{d}{dx} F^{-n}(x_o) < e^V .$$

Now, let I be an arc of S^1, of length μ, and contained in $S^1 \sim \Omega$. It is easy to see (again using the semi-conjugacy) that if n satisfies the hypotheses of the claim for some point of I, then it does for any other point of I.

Let μ_n = length of $f^n I$. Then $\mu_n = \int_I \frac{d}{dx} F^n$,

$\mu_{-n} = \int_I \frac{d}{dx} F^{-n}$. So $\mu_n + \mu_{-n} = \int_I \left(\frac{d}{dx} F^n + \frac{d}{dx} F^{-n} \right)$

$$\geq 2 \int_I \left(\frac{d}{dx} F^n \cdot \frac{d}{dx} F^{-n} \right)^{1/2} \quad \text{(by the Hölder inequality)}$$

$$> 2 \mu e^{\frac{-1}{2} V}$$

Since this holds for large n, $\mu_n + \mu_{-n} \not\to 0$, so $\sum_{-n}^{n} \mu_K = \infty$.

Hence, the segments must eventually overlap - contrary to their being contained in $S^1 \sim \Omega$.

Thus, a diffeomorphism with continuous derivative of bounded variation and with irrational rotation number is transitive. //

On the other hand, the intransitive case can occur, if the bounded variation requirement is dropped. The basic idea of the example, due again to Denjoy ([1]), is to construct a semi-conjugacy as in proposition 4, and to fill in the "holes" with an appropriate homeomorphism.

More specifically, we pick a point $x \in S^1$, and let $\{x_n\}_{n \in \mathbb{Z}}$ be its orbit under ϑ_t (t any irrational number). We construct our semi-conjugacy by specifying the inverse image of each point of ϑ_t. The inverse image of any point not in $\{x_n\}_{n \in \mathbb{Z}}$ will be a single point. The inverse image of x_n will be an interval I_n. We want its length, $\ell(I_n)$, to satisfy:

(1) $\sum_{-\infty}^{\infty} \ell(I_n) \leq 1$

(2) $\lim_{|n| \to \infty} \frac{\ell(I_n)}{\ell(I_{n+1})} = 1$

The first condition is necessary if we are to squeeze these

intervals into the circle; the second will prove necessary to make f continuously differentiable. Suppose for a moment we have found a sequence of intervals I_n, with lengths $\ell_n = \ell(I_n)$, satisfying the above two conditions.

We can then define f simply by defining f' and integrating: On $I_n = [b_n, a_n]$, we want

$$\int_{I_n} f' = \ell(I_{n+1}) \qquad (*)$$

and $f' = 1$ on the boundary of I_n. So, for $x \in I_n$, suppose

$$f'(x) = 1 + \frac{k_n(a_n-x)(x-b_n)}{\ell_n^2}, \; k_n \text{ some constant.} \qquad (**)$$

Then condition (*) above becomes

$$\int_{a_n}^{b_n} [1 + \frac{k_n(a_n-x)(x-b_n)}{\ell_n^2}]\, dx = \ell_{n+1}$$

Integration yields

$$\ell_{n+1} = \ell_n \left(1 + \frac{k_n}{6}\right)$$

or

$$k_n = \frac{6}{\ell_n} (\ell_{n+1} - \ell_n) \qquad (***)$$

Then, introducing a constant of integration so that $f(a_0) = a_1$, we can integrate (**) to get f for which

$$f(I_n) = I_{n+1}$$

and f' is continuous by definition, since by condition (2) above and (***), $\lim_{|n| \to \infty} k_n = 0$.

Now, to obtain a sequence of lengths ℓ_n as desired, we could let

$$\ell_n = \frac{L}{(n+1)(n+2)}$$

where $L \leq 1$ is the desired total length of $S^1 \sim \Omega(f)$.

Finally, to obtain a cantor set with several distinct orbits in its complement, we could pick several points $x^{(h)}$ in place of x, and substitute for each point $x_n^{(h)}$ in the orbit of $x^{(h)}$ an interval $I_n^{(h)}$ of length

$$\ell_n^{(h)} = \frac{L^{(h)}}{(n+h+1)(n+h+2)}$$

with $\sum_h L^{(h)} = L$.

We mention that Markley [1] has classified the intransitive homeomorphisms up to topological conjugacy.

We close this section with the promised discussion of the orientation-reversing case. In the theory above, the orientation-preserving hypothesis was applied to two maps — the original homeomorphism, and the conjugating map. (The rotation number is invariant only under order-preserving conjugacies: ϑ_t is conjugate to ϑ_{-t} under the map $\exp y \longmapsto \exp -y$.)

Note that an orientation-reversing homeomorphism of S^1 has exactly two fixedpoints, with two arcs in the complement of fix(f) interchanged. (To see this, consider a "lift"; then if 0 is a fixedpoint, $F(1) = -1$; thus, the function $F(x) - x$ has value 0 at 0, and -2 at 1; hence it must have value -1 somewhere between, and this represents a second fixed point on S^1. The first fixedpoint exists by virtue of the Lefschetz theorem, and the reversal of arcs follows from orientation-reversal.) In particular, since any odd iterate reverses orientation, the two fixedpoints are the only periodic points of odd period, and since arcs are reversed, $\Omega(f) = \Omega(f^2)$. But f^2 preserves orientation, so our theory applies to describe $\Omega(f^2)$. Then, once we have located the two fixedpoints of f, the rest of the periodic points of f^2 appear in pairs, which are interchanged under f, with period twice the period under f^2. Finally, we note that an orientation-reversing conjugacy can change only the sign of the rotation number (composing h with the standard orientation-reversing homeomorphism $\varphi(\exp x) = \exp(-x)$, we see that

$$\tau(\varphi\circ h\circ f\circ h^{-1}\circ\varphi^{-1}) = -\tau(h\circ f\circ h^{-1})$$

but also, since $\varphi\circ h$ preserves orientation,

$$\tau(\varphi\circ h\circ f\circ h^{-1}\circ\varphi^{-1}) = \tau(f)) .$$

Similar considerations enter in the next section.

1.2 THE GENERIC THEORY

We now shift our attention from trying to account for all possible kinds of behavior on the circle to the consideration of the "usual" behavior, which in this case also turns out to be the "stable" behavior. The theory in this section is a direct corollary of the work of Peixoto in [1]. He is concerned with the generic properties of flows on surfaces; we simply translate his methods and results to diffeomorphisms of the circle by means of cross-sections and suspensions. This has the advantage of sidestepping some of the technicalities which arise in the case of flows.

By contrast with the classical theory, where we need to consider rational and irrational, transitive and intransitive cases, the generic behavior turns out to be quite simple.

First, some terminology.

Definition: $f \in \mathrm{Diff}(S^1)$, $p \in S^1$. Then p is a hyperbolic fixedpoint of f if $f(p) = p$ and $f'(p) \neq \pm 1$.
In case $|f'(p)| < 1$, p is a sink.
If $|f'(p)| > 1$, p is a source.
A periodic point of period m is hyperbolic (resp., a sink or a source) if it has the corresponding behavior as a fixedpoint of f^m.

The results of the whole (generic) theory can then be summarized in a single

Theorem (Peixoto): A dense, open subset of $\mathrm{Diff}^r(S^1)$ $(r \geq 1)$ consists of diffeomorphisms with the following properties:
(1) $\Omega(f)$ is finite
(2) all periodic points are hyperbolic.
Any diffeomorphism satisfying (1) and (2) is structurally stable.

We remark that a finite invariant set must consist of periodic points, so (1) implies that all non-wandering points are periodic.

The observations at the end of the previous section allow us to restrict our attention to orientation-preserving maps; so we henceforth assume all diffeomorphisms preserve orientation.

The heart of this theory is the following

<u>Lemma 1</u> (Closing Lemma): $f \in \mathrm{Diff}^r(S^1)$, $p \in \Omega(f)$, then $\exists\ \tilde{f}$ arbitrarily C^r-near f with $p \in \mathrm{per}(\tilde{f})$.

Before giving a proof, we comment, first, that the case $r = 0$ is a trivial exercise, and second, that the higher-dimensional analogue of the closing lemma is known only in the case $r \leq 1$; in the latter case, it is a very difficult theorem originally due to C. Pugh ([1] and [2]; see also Pugh and Robinson [1]).

Proof of Lemma:

Either we have p recurrent — in which case for convenience we assume p recurs below itself (ie, the iterates of p have a subsequence monotonically increasing to p, relative to the orientation) — or for any neighborhood U of p, there exists $q \in U$ with $f^i q \in U$ (some $i > 0$). In this latter case, we can also assume $f^i q < q$ (if $f^i q = q$, there is no work to do). In either case, we can assume coordinates on S^1 such that p (or, respectively, q) is covered by $0 \in \mathbb{R}$.

We begin with the recurrent case.

Take U and V neighborhoods of $f^{-1}p$ with $\bar{U} \subset V$. Let φ be a C^r Urysohn function — a real-valued function on S^1 taking the value 1 on $\bar{U}$ and 0 off V. Then for ε sufficiently small and all u between 0 and 1, the map

$$f_u(x) = f(x) + \varepsilon u\, \varphi(x)$$

is a diffeomorphism of S^1 (since $f_u(x) \geq f(x)$ and

deg f_u = deg f). Also, it is clear that f_u is C^r, with $\|f-f_u\|_r \leq \varepsilon u\|\varphi\|_r \leq \varepsilon\|\varphi\|_r$.
Hence, for small ε, f_u is close to f, and clearly $f_o = f$.

Take F a lift of f to $\mathbb{R}$, and extend to a lift F_u of f_u. Then we need to show that for some $u \in [0,1]$, $(F_u)^k(0)$ is an integer.

Note that $\frac{\partial}{\partial u} F_u(x) \geq \varepsilon\ \varphi(x) \geq 0$, so that if $F_{u,k}(x)$ is any expression of the form

$$F_{u,k}(x) = F_{u_1} \circ \cdots \circ F_{u_k}(x)$$

with $u_i \leq u\ (\forall\ i)$, then

$$(F_u)^k\ (x) \geq F_{u,k}(x).$$

Now for every $y \in R$ covering a point of $\bar{U}$,

$$F_1(y) > F_o(y) .$$

Hence, F_u has a <u>minimum lift</u> δ on $\bar{U}$ - ie,

$$(\forall\ y \in \bar{U})\quad F_1(y) > F_o(y) + \delta .$$

In particular, if we have $f^i p \uparrow p$, we can pick i large so that

$$|p - f^i p| < \frac{\delta}{2}$$

and therefore

$$F_1(F_o^{i-1}(0)) > [F_o^i(0)] + 1$$

(where [z] denotes the greatest integer strictly less than z).

But then

$$(F)^i\ (0) \geq F_1(F_o^{i-1}(0)) > [F_o^i\ (0)] + 1$$

while

$(F_o)^i < [F_o^i(0)] + 1$

so that (since $(F_u)^i(x)$ varies continuously with u)

$(\exists\, u \in [0,1])\ (F_u)^i(0) = [(F_o)^i(0)] + 1 \in \mathbb{Z}.$

If on the other hand p is merely non-wandering, we have

$q \in B_{\delta/4}(p)$,

$i \in \mathbb{Z}^+ \ .\ni.\ f^i(q) \in B_{\delta/4}(p)$

or in other words,

$|f^i(q) - q| < \frac{\delta}{2}$

Conjugating by a $\frac{\delta}{4}$ - translation, we can assume $p = q$, and apply the above argument to $(F_u)^i(0)$ as before, completing the proof of the lemma. //

Our plan of attack on the main result of this section is to prove (2) generic - then (1) will follow fairly easily, as will structural stability. A crucial observation here is that hyperbolicity of a fixedpoint p is equivalent to transversal intersection at (p,p) of $gr(f)$ with the diagonal in $S^1 \times S^1$, and transversality is generic by arguments from chapter 0. However, there is a pitfall in this argument. Transversality includes two possibilities: non-intersection and transversal intersection. As we shall see below, a diffeomorphism can be easily approximated by a diffeomorphism all of whose periodic points are hyperbolic: however, in the process of approximation we may well loose any advantages we may have gained from applying the closing lemma.

In particular, given a diffeomorphism, we would like to perturb it to one with a rational rotation number, and make sure any subsequent perturbation also has rational rotation number. The initial approximation is given by the closing lemma; the stability of such a perturbation is assured by the following two lemmas.

Lemma 2: Given $f \in \mathrm{Diff}^r(S^1)$, $p \in \mathrm{per}(f)$ with period n, U a neighborhood of p, and $\varepsilon > 0$, there exists a C^r-ε-perturbation $\tilde{f}$ of f such that $\tilde{f}^n(p) = p$ and $(\tilde{f}^n)'(p) \neq 1$.

Proof

We of course need only consider the case when $(f^n)'(p) = 1$. Then we can pick a C^r-ε-perturbation α of the identity on S^1 with the properties:

(1) $\alpha | S^1 \sim U = \mathrm{id}$
(2) $\alpha^n(p) = p$
(3) $(\alpha^n)'(p) \neq 1$.

By picking U small enough, we can assume $f^i U \cap f^j U = \emptyset$ for $i \neq j$, $0 \leq i,j \leq n-1$. Then clearly $\tilde{f} = \alpha \circ f$ has the required properties. //

Lemma 3 (Stability of Hyperbolic Periodic Points): Given $f \in \mathrm{Diff}^r(S^1)$, p a hyperbolic periodic point of period n, then:

(1) there exists a closed neighborhood U of p containing no other periodic points

(2) there exists $\varepsilon > 0$ such that any C^1-ε-perturbation of f has a unique hyperbolic periodic point in U.

Proof

We prove the lemma for a fixed sink. Then all other cases can be derived by taking appropriate (positive or negative) iterates of f. (Note that a source for f is a sink for f^{-1}).

In case p is a fixed sink, we can pick U a neighborhood of p and $0 < \lambda < 1$ such that $|f'(y)| < \lambda$ for $y \in \overline{U}$. Then the length of $f(U)$ is

$$\int_U |f'(t)|dt$$

$$< \lambda \int_U t\ dt = \lambda\ \ell\ (U)$$

So $\overline{f(U)} \subset U$, and in fact $B_{1-\lambda}(f(U)) \subset U$. In particular, then, any $C^0 - (\frac{1-\lambda}{2})$ - perturbation of f will have a fixedpoint in U, and any $C^1 - (\frac{1-\lambda}{3})$ - perturbation $\tilde{f}$ of f will have $|\tilde{f}'| < \lambda$ on $\overline{U}$, for some $\tilde{\lambda} < 1$, and hence the fixedpoint will be a unique sink in U. //

The penultimate step in our "generic" treatment of diffeomorphisms of the circle is the fact that condition (2) in the theorem holds on a residual set in $\mathrm{Diff}^r(S^1)$. This is a special case of a general theorem which we shall deal with at greater length later on.

Lemma 4 ("Kupka-Smale Theorem"): Fix $k \in \mathbb{Z}^+$, and let $\mathcal{G}_k$ be the collection of diffeomorphisms of S^1 whose periodic points of period k are all hyperbolic.
Then $\mathcal{G}_k$ is open and dense in $\mathrm{Diff}^r(S^1)$ $(r \geq 1)$.

Proof

This is a straightforward application of transversality theory.

We note that $gr(f^k) = (id, f^k)$ is transverse to the diagonal, $\Delta \subset S^1 \times S^1$, if and only if its derivative does not lie in the diagonal, which in this case means $(f^k)'(y) \neq \pm 1$ if $(y, f^k y) \in \Delta$ (ie, if $y = f^k y$). Thus, $\mathcal{G}_k$ is just the set $\mathcal{G}_k = \{f \mid gr(f^k) \pitchfork \Delta\}$.

But the map $\rho^{(k)}: Diff^r(S^1) \longrightarrow C^r(S^1, S^1 \times S^1)$ given by $\rho^{(k)}(f) = gr(f^k)$ is easily seen to be a C^r-representation, so that $\mathcal{G}_k$ is open and dense by the Abraham transversality theorems. //

Corollary: If $\mathcal{G}_\infty$ is the set of diffeomorphisms all of whose periodic points are hyperbolic, then $\mathcal{G}_\infty$ is a residual subset of $Diff^r(S^1)$ $(r \geq 1)$.

Proof

Clearly, $\mathcal{G}_\infty = \bigcap_{k=1}^{\infty} \mathcal{G}_k$. //

We are now ready to complete the proof of the main theorem.

Proof of Theorem

(A) Density of (1) and (2):

Given f, we use the closing lemma (and the fact that $\Omega(f) \neq \emptyset$ for a compact manifold) to find f_1 nearby with a periodic point (ie, rational rotation number). Then we apply lemma 2 to f_1 and one of its periodic points to produce f_2 nearby with a hyperbolic periodic point. Finally, we apply the Kupka-Smale theorem to produce a nearby f_3 satisfying condition (2), noting that lemma 3 ensures that f_3 will actually have periodic points (in particular, $\tau(f_3)$ remains rational). We observe that if $\tau(f_3)$ is rational and all periodic points are hyperbolic, then f_3 also satisfies condition (1) because all points must have the same period and hyperbolic fixedpoints are (clearly) isolated.

(B) Structural Stability (and hence Openness of (1) and (2)):

Using lemma 3, we pick disjoint open sets U_j around the periodic points p_j with the property that an ε_j-perturbation of f^k has a unique fixedpoint in U_j. Then, we note that the complement of the U_j's is

$$S^1 \sim \bigcup_{j=1}^{k} U_j = I_1 \cup \cdots \cup I_k ,$$

a disjoint union of closed intervals. Since each I_i contains no fixedpoints, we can let

$$\delta_i = \min_{x \in I_i} d(x, f^k(x)) > 0 .$$

Then if $0 < \varepsilon_0 < \frac{1}{2} \min_i \delta_i$, any C^0-ε_0-perturbation of f^k will have no fixedpoints in $S^1 \sim \bigcup U_j$. Hence, any $\tilde{f}$ with $\tilde{f}^k$ C^1-ε-near f^k will have $\Omega(\tilde{f})$ conjugate to $\Omega(f)$, if $\varepsilon < \varepsilon_0, \ldots \varepsilon_k$. Finally, for structural stability, we note that sinks are perturbed to sinks, sources to sources, and for both f and $\tilde{f}$, sinks and sources alternate around S^1. In the arc between a sink and adjacent source, all points are moved toward the sink by a positive distance. Given two such arcs (I for f, $\tilde{I}$ for $\tilde{f}$) we conjugate the two maps there by picking $p \in I$, $\tilde{p} \in \tilde{I}$, picking an arbitrary homeomorphism $h:[p, f^{k-1}p] \approx [\tilde{p}, \tilde{f}^{k-1}p]$, and extending h to I by the diagram

$$\begin{array}{ccc} [f^n p,\ f^{n+k-1} p] & \xrightarrow{h} & [\tilde{f}^n \tilde{p},\ \tilde{f}^{n+k-1} \tilde{p}] \\ f^k \wr\wr & & \wr\wr\ \tilde{f}^k \\ [p,\ f^{k-1} p] & \underset{h}{\approx} & [\tilde{p},\ \tilde{f}^{k-1} \tilde{p}] \end{array}$$

It is easily checked that this consistently defines a homeomorphism $h: I \longrightarrow \tilde{I}$ conjugating $f|I$ with $\tilde{f}|\tilde{I}$, and that piecing together these definitions on all intervals between adjacent periodic points gives a conjugacy of f with $\tilde{f}$.

We close by noting that since structural stability is an open condition, conditions (1) and (2) are clearly open and dense conditions, so that the theorem is proven. //

Thus, in contrast to the complications of the most general situation (as studied in the classical theory), the "generic" situation is quite simple: a finite number of periodic points, all of the same period, alternating sinks and sources around the circle. This result is an example of complete success of the "generic" approach: classification of a dense open set of diffeomorphisms up to conjugacy, and fairly simple models for each conjugacy class.

1.3 EXPANDING MAPS

In this section, we shall temporarily divert our attention from the mainstream of our study — the generic theory of diffeomorphisms — to consider a certain generalization of this theory.

Most of the "dynamical" notions we defined in section 0.2 do not depend at all on the invertibility if the diffeomorphism in question to make sense. The notions could equally well be applied to the situation of an arbitrary differentiable map of a manifold into itself, and if analogous results could be proven, we would have a dynamical theory of C^r maps, a kind of theory of "irreversible" differentiable dynamical systems.

The beginnings of such a theory were set down by M. Shub in [1]. He formulated and proved an analogue of the Kupka-Smale theorem, and studied one large class of examples of such "irreversible" systems.

Although his examples are defined on a large family of manifolds, we shall develop his theory in the special context of the circle. As in the case of diffeomorphisms, this restriction avoids certain topological subleties, but the arguments we shall use are basically those used in higher dimensions by Shub.

<u>Definition:</u> A C^r map $f: S^1 \longrightarrow S^1$ is <u>expanding</u> iff $(\forall x \in S^1)$ $|f'(x)| > 1$.

The prototype of such maps is the endomorphism of the unit circle in the complex plane given by $e^{2\pi i t} \longrightarrow e^{2\pi i\, nt}$, $n > 1$. (Recall that the "derivative" of such a map is its derivative along the circle, which in this case is $f'(x) = n$.)

The study of expanding maps, besides its interest as a

generalization of the theory of diffeomorphisms, relates directly to that theory in two ways: it illustrates on the circle phenomena which occur even for diffeomorphisms in higher dimensions, and such maps have themselves been used (by R. Williams) in studying certain properties of diffeomorphisms.

As in previous sections, the assumption that all maps are orientation-preserving is only a slight restriction on the theory.

We begin with a few basic properties of expanding maps. Let λ be the minimum of the absolute value of the derivative of f, f an expanding map on S^1. Then $\lambda > 1$, and in the orientation-preserving case, $\lambda = \min_{x \in S^1} f'(x)$.

__Lemma 1:__ For any arc I on S^1, length $(f^n I) \geq \lambda^n$ length(I).

__Proof__

It of course suffices to show the case $n = 1$. If we let $\ell(I)$ denote length, then

$$\ell(fI) = \int_I |f'(t)|dt \geq \lambda \int_I dt = \lambda \, \ell(I) \,. \quad //$$

__Corollary:__ Expanding maps are covering maps.

__Proof__

The derivative is nowhere zero, so they are local diffeomorphisms, and are onto by the previous lemma. //

__Lemma 2:__ S^1 has a residual set of points with dense orbit under f.

Proof

Take $\{U_i\}_{i \in \mathbb{Z}_+}$ a countable basis for the topology on S^1 (say, $B_{\frac{1}{n}}(\frac{p}{q}) = [\frac{p}{q} - \frac{1}{n}, \frac{p}{q} + \frac{1}{n}]$, n, p, q $\in \mathbb{Z}$). Then let

$V_i = \{x \in S^1 \mid (\exists n > 0)\ f^n x \in U_i\}$.

V_i is the set of points which will, at some future time, land in U_i. Clearly, V_i is open ($V_i = \bigcup_{k \geq 0} f^{-k}[U_i]$) and dense by Lemma 1 (there can be no interval in $S^1 \sim V_i$, since every interval of positive length eventually fills up the circle). But then $V_\infty = \bigcap_{i \in \mathbb{Z}_+} V_i$ is clearly residual, and any point $x \in V_\infty$ has dense orbit since $O^+(x)$ intersects every element of a basis. //

Lemma 3: Periodic points of f are dense in S^1.

Proof

Given a non-trivial interval $I \subset S^1$, we need to show that $I \cap \text{per}(f) \neq \emptyset$. We can assume I closed. By Lemma 1, $\exists\, n \in \mathbb{Z}_+$ such that $f^n I = S^1$; we shall show that I contains a fixedpoint of f^{2n}.

Since $f^n I = S^1$, I at least double-covers itself under f^{2n}, in the sense that each point has at least two inverse images in I. In particular, some component of $f^{-2n}[I]$ is contained in I; pick such a component and call it I_1. By induction, we take I_k an interval, contained in I_{k-1}, mapping onto I_{k-1} under f^{2n}. These form a sequence of nested, closed intervals, and by lemma 1, their lengths tend to 0, so that their intersection is a point, p, in I, which clearly is a fixedpoint of f^{2n}. //

Corollary: $\Omega(f) = S^1$.

Proof

This is clear since $\overline{\mathrm{per}(f)} \subset \Omega(f)$. //

Reflecting for a moment, we see that several features of the situation here are very different from the picture for diffeomorphisms. Whereas $\Omega(f)$ was generically finite for diffeomorphisms, here $\Omega(f)$ is the whole circle; whereas sinks and sources alternated for diffeomorphisms, here all periodic points are sources, and furthermore, whereas a source is always isolated in $\Omega(f)$ for f a diffeomorphism, here we have a dense set of sources. We shall see later that the situation of $\Omega(f)$ being the whole manifold can arise for f a diffeomorphism in higher dimensions. (Indeed, it arises in the "transitive" case on S^1, but this can be perturbated away.) The other two phenomena are specific to endomorphisms.

We would like to show that the phenomena mentioned above are stable. Before we state Shub's theorem on this point, however, we would like to define the degree of a map $S^1 \longrightarrow S^1$. This can, of course, be defined in terms of homology, but we take a more analytic and pedestrian approach.

Definition: If $\alpha \in C^0(S^1, S^1)$, then a lift of α is any continuous map

$\bar{\alpha} : \mathbb{R} \longrightarrow \mathbb{R}$

such that $\exp \bar{\alpha}\, x = \alpha \exp x \quad (\forall\, x \in \mathbb{R})$.

We note that the "lifts" defined in section 1.1 are special cases of the above.

We claim:

Lemma 4: $\frac{1}{n}(\bar{\alpha}(x+n) - \bar{\alpha}(x))$ is an integer depending only on α.

Proof

(1) $\bar{\alpha}(x+n) - \bar{\alpha}(x)$ is an integer because $\exp \bar{\alpha}(x+n) = \alpha \exp(x+n) = \alpha \exp(x) = \exp \bar{\alpha}(x)$.

(2) Hence, by continuity, $\bar{\alpha}(x+n) - \bar{\alpha}(x)$ is independent of x.

(3) If $\bar{\alpha}$ and $\tilde{\alpha}$ are two lifts of α, then $\exp \circ \bar{\alpha} = \alpha \circ \exp = \exp \circ \tilde{\alpha}$, so $\bar{\alpha}(x) = \tilde{\alpha}(x) + k$, and again k is clearly independent of x.

(4) Finally, $\frac{1}{n}(\bar{\alpha}(x+n) - \bar{\alpha}(x))$ is independent of n, since

$$\frac{1}{n}(\bar{\alpha}(x+n) - \bar{\alpha}(x)) = \frac{1}{n}\sum_{i=1}^{n}[\bar{\alpha}(x+i) - \bar{\alpha}(x+i-1)]$$

$$= \frac{1}{n}\sum_{i=1}^{n}[\bar{\alpha}(x+1) - \bar{\alpha}(x)] \quad \text{by (2))}$$

$$= \bar{\alpha}(x+1) - \bar{\alpha}(x) \;. \quad //$$

Now, invoking this lemma, we can state a

Definition: For any $\alpha \in C^0(S^1, S^1)$, set the degree of α to be

$$\deg(\alpha) = \frac{1}{n}(\bar{\alpha}(x+n) - \bar{\alpha}(x))$$

for any $n \in \mathbb{Z}$ and any lift $\bar{\alpha}$ of α.

Shub's main theorem concerning expanding maps then states

Theorem (Shub): If f and g are expanding maps of the same degree, they are topologically conjugate.

We give here the proof which Shub gave in [1], which generalizes to higher dimensions. In Appendix I, we give two alternate proofs in the case of the circle, using techniques that can be used to study other circle endomorphisms. Our present proof follows from

<u>Lemma 5</u>: Suppose f is an expanding map and g is any continuous map, of the same degree as f. Then given lifts $\bar{f}$ and $\bar{g}$ of f and g (resp.), there exists a unique $\bar{\alpha}$, lifting a continuous map $\alpha : S^1 \longrightarrow S^1$ of degree one, with

$$\begin{array}{ccc} \mathbb{R} & \xrightarrow{\ \bar{f}\ } & \mathbb{R} \\ {\scriptstyle \bar{\alpha}}\downarrow & & \downarrow{\scriptstyle \bar{\alpha}} \\ \mathbb{R} & \xrightarrow[\ \bar{g}\ ]{} & \mathbb{R} \end{array}$$

commuting.

<u>Proof</u>

Let V be the space of lifts of maps of degree one, with the sup metric $\left(\mathcal{D}(\bar{\alpha}, \bar{\beta}) = \max_{x \in \mathbb{R}} (d(\bar{\alpha}x, \bar{\beta}x)) \right)$

1. V is a complete metric space: The sup metric is clearly a metric if finite everywhere. But a lift of a map of degree one differs from the identity by a periodic map of period one, and so in fact

$$\mathcal{D}(\bar{\alpha}, \bar{\beta}) = \max_{0 \le x \le 1} d(\bar{\alpha}x, \bar{\beta}x) .$$

V is complete because the sup-norm topology is that of uniform convergence, and a uniform limit of lifts of degree-one maps is itself a lift: if $\bar{\alpha}_i \longrightarrow \bar{\alpha}$,

$$\begin{aligned}\exp \bar{\alpha}\,(x) &= \lim \exp \bar{\alpha}_i(x) \\ &= \lim \alpha_i\,(\exp x) = \alpha(\exp x)\ .\end{aligned}$$

2. Define the operator $T\colon V \longrightarrow C^0(\mathbb{R})$ by $T\bar{\alpha} = \bar{f}^{-1}\,\bar{\alpha}\,\bar{g}$. T maps V into V, since

$$\begin{aligned}\bar{f}\circ T(\bar{\alpha})(x+n) &= \bar{f}\,\bar{f}^{-1}\,\bar{\alpha}\,\bar{g}\,(x+n) \\ &= \bar{\alpha}\,(\bar{g}x + n \deg g) \\ &= \bar{\alpha}\,\bar{g}\,x + n \deg g \\ &= \bar{f}(T(\bar{\alpha})(x)) + [T(\bar{\alpha})(x+n) - T(\bar{\alpha})x] \deg f \\ &= \bar{\alpha}\,\bar{g}(x) + [T(\bar{\alpha})(x+n) - T(\bar{\alpha})(x)] \deg f\end{aligned}$$

ie, $T(\bar{\alpha})(x+n) - T(\bar{\alpha})(x) = n$ so that $T(\bar{\alpha}) \in V$.

Also, T is a contraction on V, because

$$\begin{aligned}\mathcal{D}\,(T(\bar{\alpha}),\ T(\bar{\beta})) &= \sup_{x\in\mathbb{R}}\ d(T(\bar{\alpha})(x),\ T(\bar{\beta})(x)) \\ &= \sup_{x\in\mathbb{R}}\ d(\bar{f}^{-1}\,\bar{\alpha}\,\bar{g}(x),\ \bar{f}^{-1}\,\bar{\beta}\,\bar{g}(x)) \\ &= \sup_{y=\bar{g}(x)\in\mathbb{R}}\ d(\bar{f}^{-1}\,\bar{\alpha}\,(y),\ \bar{f}^{-1}\,\bar{\beta}(y)) \\ &\le \sup_{y\in\mathbb{R}}\ \lambda^{-n}\, d(\bar{\alpha}(y),\ \bar{\beta}(y)) \\ &= \lambda^{-n}\,\mathcal{D}\,(\bar{\alpha},\ \bar{\beta}).\end{aligned}$$

Hence, T has a unique fixedpoint, $\bar{\alpha}$ in V. This means $T(\bar{\alpha}) = \bar{\alpha}$, or $\bar{f}^{-1}\,\bar{\alpha}\,\bar{g} = \bar{\alpha}$, or $\bar{\alpha}\,\bar{g} = \bar{f}\,\bar{\alpha}$, and projecting to S^1 (since $\bar{\alpha}$ is a lift of α), we have

$$\begin{array}{ccc} S^1 & \xrightarrow{\ f\ } & S^1 \\ {\scriptstyle\alpha}\downarrow & & \downarrow{\scriptstyle\alpha} \\ S^1 & \xrightarrow[\ g\ ]{} & S^1 \end{array}$$

commuting. //

Proof of Theorem

Finally, if f and g are both expanding, we apply lemma 5 thrice. We pick $\bar{f}$, $\bar{g}$ lifts of f and g.

1. f is expanding, g continuous, and $\deg f = \deg g$, so we have $\bar{\alpha}$ conjugating $\bar{f}$ and $\bar{g}$.

2. g is expanding, f continuous, so similarly we have $\bar{\beta}$ conjugating $\bar{g}$ and $\bar{f}$.

3. Putting these together, we have

$$\begin{array}{ccc} R & \xrightarrow{\bar{f}} & R \\ \bar{\alpha}\downarrow & & \downarrow\bar{\alpha} \\ R & \xrightarrow{\bar{g}} & R \\ \bar{\beta}\downarrow & & \downarrow\bar{\beta} \\ R & \xrightarrow{\bar{f}} & R \end{array}$$

but we also have

$$\begin{array}{ccc} R & \xrightarrow{\bar{f}} & R \\ \text{id}\downarrow & & \downarrow\text{id} \\ R & \xrightarrow{\bar{f}} & R \end{array}$$

and Lemma 5 asserts uniqueness, so that $\bar{\beta}\circ\bar{\alpha} = \text{id}$. Similarly, $\bar{\alpha}\circ\bar{\beta} = \text{id}$, so $\bar{\alpha} = \bar{\beta}^{-1}$, and thus α is a homeomorphism. //

As an immediate consequence, we have:

Corollary: Expanding maps are structurally stable.

Proof

Any two sufficiently C^0 -close maps are of the same degree, and any map sufficiently C^1 -close to an expanding map is itself expanding. Thus, the theorem applies, and we are done.//

CHAPTER 2

PERIODIC POINTS

In this chapter, we will be concerned with developing a "generic" picture of the local behavior of a dynamical system around its periodic points. We will begin by defining the appropriate generalization of the "sinks" and "sources" of chapter 1, and proving Hartman's theorem, which gives a fairly complete topological picture of the behavior of a system near a hyperbolic fixedpoint. We will then consider the perturbation theory of hyperbolic periodic points, first showing their local stability, then proving their genericity, in a reduced version of the Kupka-Smale theorem. Finally, we will consider Morse-Smale systems, a class of examples in which the whole non-wandering set consists of hyperbolic periodic points.

2.0 LIPSCHITZ FUNCTIONS

We collect here the technical data concerning Lipschitz functions which we will need in proving Hartman's theorem.

Definition: X a metric space, then a function $f:X \longrightarrow X$ is Lipschitz if there exists a constant $K \in \mathbb{R}$ such that for any points $x, y \in X$,

$$d(f(x), f(y)) \leq K\, d(x,y) .$$

The least constant K for which the above estimate holds

universally is called the Lipschitz constant of f, denoted $\mathcal{L}ip(f)$.

As an example, we remark that any bounded linear operator L on a Banach space is Lipschitz, with $\mathcal{L}ip(L) = \|L\|$. In general, the Lipschitz condition can be regarded as a type of smoothness condition somewhere between (uniform) continuity and differentiability.

The best-known theorem about Lipschitz maps is

Lemma 1 (Contraction Lemma): If X is a complete metric space and f is a Lipschitz map with $\mathcal{L}ip(f) < 1$, then f has a unique fixedpoint expressible as $\lim_{n \to \infty} f^n x$ for any $x \in X$.

This follows immediately from the observations: (1) that the sequence $\{f^n x\}_{n=0}^{\infty}$ is Cauchy, for any $x \in X$, (2) that any two such sequences are asymptotic, and (3) that any limit point of such a sequence is fixed. //

We use the contraction lemma to prove the following modification of the classical inverse function theorem:

Proposition (Lipschitz Inverse Function Theorem): Suppose X is a Banach space and $L: X \longrightarrow X$ is an invertible linear operator. If ϕ is a Lipschitz map with

$$\mathcal{L}ip(\phi) < \|L^{-1}\|^{-1},$$

then $(L + \phi)$ is invertible, with Lipschitz inverse, and

$$\mathcal{L}ip[(L + \phi)^{-1}] \leq \frac{1}{\|L^{-1}\|^{-1} - \mathcal{L}ip\,\phi}$$

Proof

(1) We first show $L + \phi$ injective by demonstrating the bound above on $\mathcal{L}ip[(L + \phi)^{-1}]$:

$$\|(L + \phi)x - (L + \phi)y\| = \|L(x-y) + \phi x - \phi y\|$$

$$\geq [\ \|L^{-1}\|^{-1} - \mathcal{L}ip(\phi)]\ \|x - y\| .$$

(2) It remains, then, to show $(L + \phi)$ surjective (so that $(L + \phi)^{-1}$ is defined everywhere on X)—ie, given $y \in X$, we wish to solve the equation

$$(L + \phi)(x) = y \tag{1}$$

for $x \in X$. Now, for $\phi = 0$, the equation becomes

$$L(x) = y \tag{2}$$

which has a solution $x_o = L^{-1}y$. Equation (1) can then be restated

$$(L + \phi)(x_o + \delta) = y \tag{3}$$

which can be expanded to

$$Lx_o + L\delta + \phi(x_o + \delta) = L x_o$$

or

$$L\delta + \phi(x_o + \delta) = 0$$

which is (since L is invertible)

$$\delta + L^{-1}\phi(x_o + \delta) = 0$$

or

$$\delta = -L^{-1}\phi(x_o + \delta) . \tag{4}$$

Now, consider the operator

$$\mu(\delta) = -L^{-1}(\phi(x_o + \delta)) .$$

We note that μ is Lipschitz, with

$$\mathcal{L}ip(\mu) \leq \|L^{-1}\|\ \mathcal{L}ip\ \phi$$

which is less than 1 by hypothesis, and so has a fixedpoint by the contraction mapping lemma. This fixedpoint clearly solves (4), and hence (1). //

A more general version of this proposition, involving the case of a Lipschitz map between two different metric spaces, is proven in §1 of Hirsch and Pugh [1].

2.1 LOCAL BEHAVIOR OF PERIODIC POINTS

For reasons that will become clear in succeeding chapters, we shall treat as much of the theory as possible in a Banach-space context.

Let E be a Banach space, with norm denoted $\|\cdot\|$.

Definition: An invertible linear map $L:E \longrightarrow E$ is *hyperbolic* if and only if the spectrum of L is disjoint from the unit circle.

Definition: Let V be a neighborhood of $0 \in E$, and $f:V \longrightarrow E$ a diffeomorphism onto its image. Then 0 is a *hyperbolic fixedpoint* of f if and only if:

(1) $f(0) = 0$

and (2) $(\mathcal{D} f)_0 : E \longrightarrow E$ is hyperbolic.

We remark, in relation to chapter 1, that a hyperbolic linear map $L:R \longrightarrow R$ is just multiplication by some real number other than 0 or ± 1.

The dynamic properties of a hyperbolic linear map are made clearer by the next lemma, whose conclusion could be taken as an equivalent definition of hyperbolicity.

Lemma: If $L:E \longrightarrow E$ is a hyperbolic linear isomorphism, then there exists an equivalent norm, $\|\cdot\|_2$, on E and a direct-sum decomposition of E satisfying

(1) $E = E_s \oplus E_u$

(2) E_s and E_u are each invariant under L

(3) If $L_s = L|E_s$, $L_u = L|E_u$, then L_s and L_u^{-1} each have operator norm less than 1, with respect to $\|\cdot\|_2$.

Proof (Palis, [1])

The first two statements follow from basic spectral theory in Banach spaces (see, for instance, Riesz-Nagy, [1] for further details). One shows that

$$P = \frac{-1}{2\pi i} \int_{S^1} R_\gamma \, d\gamma$$

is a projection map, where R_γ is the resolvent of L. Then one defines

$$E_s = \text{image } (P) = \{z \mid P(z) = z\}$$

$$E_u = \text{kernel } (P) = \{z \mid P(z) = 0\} .$$

To show that last statement, we note that by definition, the spectral radius of L_S is $r_s < 1$, so there exists a real number, a, between r_s and 1, and a certain (positive) iterate of L_s such that

$$\| L_s^p \| < a^p .$$

Now, the numerical series

$$\sum_{n=o}^{\infty} a^{-pn} \| L^{pn} \|$$

converges (it is bounded by a geometric series) and so the series

$$\sum_{n=o}^{\infty} a^{-n} \| L_s^n \| ,$$

being a sum of p series like the previous one, itself

converges.

Given this fact, we can set

$$\| v \|_2 = \sum_{n=0}^{\infty} a^{-n} \| L^n_s v \|$$

for any $v \in E_s$.

Similarly, we can define

$$\| v \|_2 = \sum_{n=0}^{\infty} a^{-n} \| L_u^{-n} v \|$$

for any $v \in E_u$, and extend the norm to $E = E_s \oplus E_u$ by letting

$$(\| v_s \oplus v_u \|_2)^2 = (\| v_s \|_2)^2 + (\| v_u \|_2)^2$$

or

$$\| v_s \oplus v_u \|_2 = \max \; (\| v_s \|_2 , \| v_u \|_2)$$

Then it is clear that

$$\| L_s \|_2 < a$$

$$\| L_u^{-1} \|_2 < a .$$

Finally, to show $\| \cdot \|_2$ equivalent to the old norm, $\| \cdot \|$, we simply note that

$$\sum_{1}^{\infty} a^{-n} \| L_s^n \| \| v \| \leq \| v \|_2 \leq \sum_{0}^{\infty} a^{-n} \| L_s^n \| \| v \|$$

for any $v \in E_s$, and similar inequalities hold for $v \in E_u$. //

This lemma gives us a reasonable picture of the dynamical behavior of a general hyperbolic linear map: the space divides into a "contracting" and "expanding" part, and points in neither must, by linearity, be forward-asymptotic to E_u and backward-asymptotic to E_s (see fig. 1).

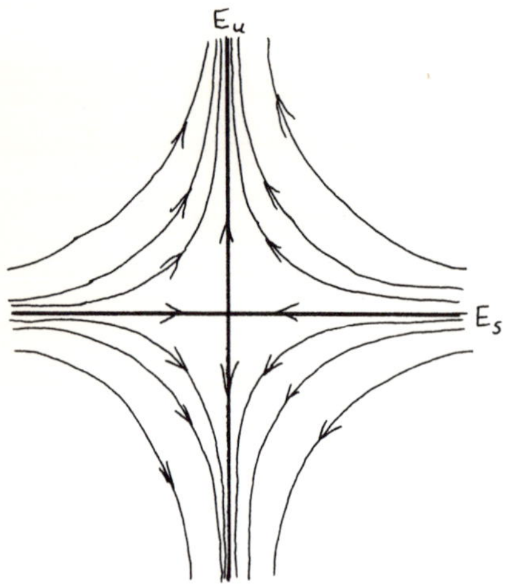

Figure 1. Dynamic Structure of a Hyperbolic Linear Map.

We note that, in special cases, one of the two subspaces of the splitting might be trivial. We distinguish these cases when considering hyperbolic fixedpoints by calling 0

(1) a <u>source</u> if $E = E_u$ for $\mathcal{D}f_0$
(2) a <u>sink</u> if $E = E_s$ for $\mathcal{D}f_0$
(3) a <u>saddle</u> if both E_u and E_s are non-trivial.

We also speak of <u>hyperbolic periodic points</u> by regarding their properties as fixedpoints of an appropriate iterate of the map. (We will often prove results for fixedpoints and apply them to periodic points; the translation can always be made in the above manner.)

The basic result concerning the local behavior of hyperbolic fixedpoints is that the map in a neighborhood of a hyperbolic point is topologically conjugate to the derivative at the fixedpoint. This result was first proven by Hartman ([1]) and Grobman ([1]) in the finite-dimensional case, and is now generally known as Hartman's theorem. The proof we give here, which covers the infinite-dimensional cases as well, is due independently to Palis ([1]) and Pugh ([3]), building on an idea due to Moser ([1]).

The first major step in the proof is the following theorem, which will also prove useful in the next section.

Theorem 1: For any hyperbolic linear automorphism L of a Banach space E, there exists a constant $\varepsilon > 0$ such that for any continuous, bounded map on E with Lipschitz constant

$$\mathcal{L}ip\,(\phi) < \varepsilon$$

there exists a unique map $h_\phi \in C^0_b(E)$ satisfying

$$(id + h_\phi) \circ L = (L + \phi) \circ (id + h_\phi) .$$

Furthermore, for perhaps a smaller ε, h_ϕ is a homeomorphism.

The theorem will follow easily from

Lemma: If L is as above, then the functional equation

$$(L + \psi) \circ (id + h) = (id + h) \circ (L + \phi) \tag{A}$$

has a unique solution $h_{\psi\phi}$ in $C^0_b(E)$ provided $\psi, \phi \in C^0_b$ have sufficiently small Lipschitz constants, the bound depending only on L.

Proof of Lemma

(1) If $\mathcal{L}ip(\phi) < \| L^{-1} \|^{-1}$, then we already have $L + \phi$ invertible (by the Lipschitz inverse function theorem).

Equation (A) expands to

$$L + Lh + \psi \circ (id + h) = L + \phi + h \circ (L + \phi)$$

or

$Lh - h(L + \phi) = \phi - \psi \circ (id + h).$

Now, call the left side of (B) $\mathcal{L}_\phi(h)$, and let $\mathcal{L}^*_\phi(v) = L \circ v \circ (L + \phi)^{-1}$. Then $\mathcal{L}_\phi = (\mathcal{L}^*_\phi - id) \circ \alpha_{(L+\phi)}$ (where, recall, $\alpha_k(v) = v \circ k$).

(2) We <u>claim:</u> $\mathcal{L}^*_\phi$ is hyperbolic .

We simply produce a splitting as in the first lemma of this section. Note that $C^0_b(E, E)$ splits as a direct sum

$$C_b{}^0(E,E) = C_b{}^0(E,E_s) \oplus C_b{}^0(E,E_w)$$

and letting $\mathcal{L}^s_\phi = \mathcal{L}^*_\phi \mid C_b{}^0(E,E_s)$

$$\mathcal{L}^u_\phi = {}^*_\phi \mid C_b{}^0(E,E_u)$$

which can be rewritten

$$\mathcal{L}^s_\phi(v_s) = L_s \circ v_s \circ (L + \phi)^{-1}$$

$$\mathcal{L}^u_\phi(v_u) = L_u \circ v_u \circ (L + \phi)^{-1}$$

so that

$$\| \mathcal{L}^s_\phi \| = \| L_s \| \leq \lambda < 1$$

and since

$$(\mathcal{L}^u_\phi)^{-1}(v) = (L_u{}^{-1}) \circ v \circ (L + \phi)$$

we have also

$$\| (\mathcal{L}^u_\phi)^{-1} \| = \| L_u{}^{-1} \| \leq \lambda < 1 .$$

(3) Next, we <u>claim:</u> $\mathcal{L}_\phi$ is invertible, and

$$\frac{1}{\|L\| + 1} \leq \|\mathcal{L}^{-1}_\phi\| \leq \frac{1}{\min(1-\lambda, \lambda^{-1}-1)}$$

The invertibility of $\mathcal{L}_\phi$ follows from the classical result: if H is linear and hyperbolic, then $(H - id)$ is invertible.

Since $\left\| \alpha_{(L+\phi)^{-1}} \right\| = 1$, it will suffice to find $\left\| (\mathcal{L}_{\phi}^{*} - \mathrm{id})^{-1} \right\|$, to establish the above estimates. We let $H = \mathcal{L}_{\phi}^{*}$. Then

$$\begin{aligned}(1 + \lambda) \|x_s\| &\geq \|x_s\| + \|H_s x_s\| \\ &\geq \|x_s - H_s x_s\| \geq \|x_s\| - \|H_s x_s\| \\ &\geq (1 - \lambda) \|x_s\|\end{aligned}$$

and

$$\begin{aligned}(\|L_u\| + 1) \|x_u\| &\geq \|H_u x_u\| + \|x_u\| \\ &\geq \|H_u x_u - x_u\| \geq \|H_u x_u\| - \|x_u\| \\ &\geq (\lambda^{-1} - 1) \|x_u\|.\end{aligned}$$

Thus taking reciprocals, maximizing, and noting that $\|L\| = \|L_u\|$, we have

$$\frac{1}{\max(\|L\| + 1, \lambda + 1)} \leq \left\| (H - \mathrm{id})^{-1} \right\| \leq \frac{1}{\min(1 - \lambda, \lambda^{-1} - 1)}$$

(4) Now, equation (B) reduces to

$$h = \mathcal{L}_{\phi}^{-1} (\phi - \psi \circ (\mathrm{id} + h)) \tag{C}$$

Letting $\mu(h)$ represent the right-hand side of (C), we <u>claim:</u> μ is a contraction if $\mathcal{L}\mathrm{ip} < \left\| \mathcal{L}_{\phi}^{-1} \right\|^{-1}$.

$$\begin{aligned}\|\mu(v) - \mu(w)\| &= \left\| \mathcal{L}_{\phi}^{-1}(\phi - \psi(\mathrm{id} + v)) - \mathcal{L}_{\phi}^{-1} (\phi - \psi(\mathrm{id} + w)) \right\| \\ &= \left\| \mathcal{L}_{\phi}^{-1}(\psi(\mathrm{id} + w) - \psi(\mathrm{id} + v)) \right\| \\ &\leq \left\| \mathcal{L}_{\phi}^{-1} \right\| \mathcal{L}\mathrm{ip}\,\psi \, \|w - v\| \\ &< \|v - w\| \text{ if } \mathcal{L}\mathrm{ip}\,\psi < \left\| \mathcal{L}_{\phi}^{-1} \right\|^{-1}.\end{aligned}$$

(5) Thus, if in particular

$$\mathcal{L}\mathrm{ip}\,\psi < \frac{1}{\max(1 + \lambda, \lambda^{-1} + 1)}$$

then μ has a unique fixedpoint, $h_{\psi\phi}$, which solves (C), hence (B), hence (A). //

We can now give the

<u>Proof of Theorem 1</u>

(1) Setting $\psi = 0$ in the lemma, we get a continuous map h_ϕ making the following commute

$$\begin{array}{ccc} E & \xrightarrow{\ L\ } & E \\ {\scriptstyle id+h_\phi}\downarrow & & \downarrow{\scriptstyle id\,+\,h_\phi} \\ E & \xrightarrow{\ L+\phi\ } & E \end{array}$$

(2) On the other hand, setting $\phi = 0$ in the lemma and setting "ψ" in the lemma equal to "ϕ" in the statement of the theorem, we get a unique map h'_ϕ making the following commute:

$$\begin{array}{ccc} E & \xrightarrow{\ L+\phi\ } & E \\ {\scriptstyle id+h'_\phi}\downarrow & & \downarrow{\scriptstyle id+h'_\phi} \\ E & \xrightarrow[\ L\ ]{} & E \end{array}$$

(3) Thus, we can compose the above two blocks in either order to get $(id + h'_\phi) \circ (id + h_\phi)$ conjugating L with itself and $(id + h_\phi) \circ (id + h'_\phi)$ conjugating $L + \phi$ with itself. But then by similar uniqueness considerations to those above, we must have

$$(id + h'_\phi) \circ (id + h_\phi) = id$$

$$(id + h_\phi) \circ (id + h'_\phi) = id$$

so that each is a homeomorphism.

(4) We close by recalling that the only bounds assumed in the above argument were

$$\mathcal{L}ip\ \phi < \|L^{-1}\|^{-1}$$

$$\mathcal{L}ip\ \psi < \frac{1}{\max(1+\lambda, \lambda^{-1}+1)}$$

thus verifying our claim that these bounds depend only on L.//

Theorem 1 is our basic tool in proving Hartman's theorem: we need only verify the hypotheses of theorem 1 in our situation. This is embodied in a

<u>Lemma</u>: Let V represent a neighborhood of 0 in Banach space E, and let $\phi: V \longrightarrow E$ be a C^1 map satisfying

(1) $\phi(0) = 0$
(2) $(D\phi)_0 = 0$.

Then for every $\varepsilon > 0$, there exists an open subset W of V, containing 0, and a map $\tilde{\phi}: E \longrightarrow E$ satisfying

(1) $\tilde{\phi}|W = \phi|W$
(2) $\tilde{\phi}$ is bounded with

$$\mathcal{L}ip\ (\tilde{\phi}) < \varepsilon\ .$$

This lemma says, in effect, that in a small neighborhood of a hyperbolic fixedpoint, a diffeomorphism can be regarded as a small Lipschitz perturbation of its derivative at the fixedpoint, and hence allows us to deduce Hartman's theorem from theorem 1.

<u>Proof</u> (Palis, [1])

We start by taking α to be a smooth (C^∞) "bump" function $\alpha: \mathbb{R} \longrightarrow [0,1]$ satisfying

(1) $\alpha(t) = \begin{cases} 1 & \text{if } |t| < \frac{1}{2} \\ 0 & \text{if } |t| \geq 1 \end{cases}$

(2) $|D\alpha| \leq 2$

The existence of such functions is a standard result in

differential topology (see for example Munkres [1], p. 6)

Now, by the continuity of $D\phi$, there is a radius $\rho > 0$ such that for z in the (open) ρ-ball about 0, $\|(D\phi)_z\| < \varepsilon/3$. We observe that then also $\|\phi(z)\| < \varepsilon/3$ for $\|z\| < \rho$.

Define the map $\tilde{\phi}$ by

$$\tilde{\phi}(z) = \begin{cases} \alpha(\frac{1}{\rho}\|z\|)\,\phi(z) & \text{if } \|z\| \leq \rho \\ 0 & \text{if } \|z\| > \rho \end{cases}$$

Note that we make no claims of differentiability for $\tilde{\phi}$ (see Bonic and Frampton, [1], for reasons why this might even be impossible). However, $\tilde{\phi}$ does satisfy:

(1) $\tilde{\phi}(z) = \phi(z)$ for $\|z\| \leq \frac{\rho}{2}$

(2) $\tilde{\phi}$ is bounded (since ϕ is bounded on a small ball).

(3) $\mathcal{L}ip(\tilde{\phi}) \leq \sup \|D\phi\|$

$$\leq \frac{2\varepsilon}{3} < \varepsilon .$$

Thus, the function defined is the desired one. //

As an almost immediate consequence of this lemma and theorem 1, we can state

<u>Theorem 2</u> (Hartman's Theorem): Let V denote a neighborhood of 0 in a Banach space E, and let

$$f: V \longrightarrow E$$

be a C^1 diffeomorphism with its image, such that 0 is a hyperbolic fixedpoint of f.

Then there exist a neighborhood U of 0 contained in V and a homeomorphism h of U onto a neighborhood of 0 in E such that

$$h\, L(z) = f\, h(z)$$

for all z in U, where

$L = (D\,f)_0$.

Proof

We apply the above lemma to the locally defined map

$\phi(z) = f(z) - L(z)$

so as to write

$\tilde{f} = L + \tilde{\phi}$

where $\tilde{f}$ extends f from some ball about 0. Then, by taking that ball small enough, we can insure that $\tilde{\phi}$ satisfies the hypotheses of theorem 1, and thus that there is a unique bounded homeomorphism h such that

$h \circ L = \tilde{f} \circ h$.

Restricting to the ball on which $\tilde{f}h = fh$, we have the theorem. //

2.2 PERTURBATIONS OF PERIODIC POINTS

This section provides in some sense the rationale for our study of hyperbolic periodic points in the last section. Recall that our overall concern in dynamical systems theory is the identification and description of generic and stable phenomena. In the study of periodic points, these two types of questions both have satisfactory answers, as it will turn out that hyperbolic points are both stable and generic. We prove these facts here.

Our description of the behavior of hyperbolic fixedpoints in the last section was purely local, in the sense that it was confined to a small neighborhood of the point in question. The type of stability this description enjoys will accordingly be local, too. The basic fact is

<u>Theorem 3</u> (Local Stability of Hyperbolic Fixedpoints): Suppose we have the situation of Hartman's theorem — that is,

$f: V \longrightarrow E$

a C^1 embedding of a neighborhood of the origin into a Banach space E with the origin, 0, a hyperbolic fixedpoint.

Then there is a neighborhood N of f in $C^1(V,E)$ such that every g in N is locally conjugate to f — that is, there will be neighborhoods U and W of 0 in E and a homeomorphism

$h: W \longrightarrow U$

such that the diagram

$$\begin{array}{ccc} V \supset W & \xrightarrow{\quad f \quad} & E \supset W \\ h \downarrow & & \downarrow h \\ U & \xrightarrow[\quad g \quad]{} & E \supset U \end{array}$$

commutes.

Furthermore, $h(0)$ will be a hyperbolic fixedpoint of g.

<u>Proof</u>

This is essentially a corollary of the techniques of section 1:

We begin with a classical observation.

<u>Claim</u>: Hyperbolic linear automorphisms of E are an open subset of $L(E,E)$, the space of linear maps.

This follows immediately from several classical results: semi-continuity of the spectrum under perturbation, compactness of S^1, and openness of invertibility for linear maps. The reader should be able to fill in the proof himself, or consult Riesz-Nagy [1] for more details.

Next, we apply the lemma immediately preceding theorem 2 to get, locally,

$f = (D\,f)_o + \phi$

with $\mathcal{L}ip\ \phi$ very small. Also g C^1-near f means in particular that (perhaps on a smaller set)

$g = f + \psi$

with $\mathcal{L}ip\ \psi$ very small.

But then we can apply theorem 1 to

$g = (D\,f)_o + \phi + \psi$

and with $\mathcal{L}ip\ (\phi + \psi)$ small, we get $(D\,f)_o \sim g$, locally. But since $f \sim (D\,f)_o$ locally by theorem 2, we have $f \sim g$, locally. In particular, this means that g will have a unique fixedpoint near the origin, and then the openness of hyperbolicity (claim above) guarantees that this fixedpoint will be hyperbolic (since $(D\,g)_{h(o)}$ is near $(D\,f)_o$, as a linear operator). //

We remark that of course the stability of hyperbolic fixedpoints implies their inclusion in any "generic" description of orbit structure. We now show a kind of converse — that as far as periodic points go, the hyperbolic description suffices.

<u>Theorem 4</u> ("Kupka-Smale Theorem"): Let M be a smooth compact manifold.

Let $H_n = \{f \in \mathrm{Diff}(M) \mid f^n$ has only hyperbolic fixedpoints$\}$

Then H_n is (C^1) open and (C^r) dense, so that

$H_\infty = \bigcap_{n=1}^{\infty} H_n = \{f | f$ has only hyperbolic periodic points$\}$ is residual in $\text{Diff}^r(M)$.

We remark that this is really only part of what is usually referred to as the Kupka-Smale Theorem; the rest of the theorem deals with transversality of invariant manifolds, which we shall touch on in the next section, and treat more fully in Appendix II.

Our theorem is proved in a sequence of lemmas. The reader is urged to compare the proofs of lemmas 2 - 4 of Chapter 1 (pp. 55-66), of which of course the argument here is a generalization.

We begin with a partial globalization of theorem 3.

Lemma (Openness of H_n): If $f \in H_n$, then there is a neighborhood U of f in $\text{Diff}^1(M)$ such that every g in U is in H_n.

Proof

Since we can insure that g^n is near f^n by making g near f, it suffices to prove the lemma for $n = 1$.

Note that, since hyperbolic fixedpoints are isolated, $f \in H_1$ means in particular that $\text{fix}(f)$ is a finite set, say $\text{fix}(f) = \{x_1, \ldots, x_k\}$. Let $V_1 \ldots V_k$ be open balls about $x_1 \ldots x_k$ and $U_1, \ldots, U_k$ neighborhoods of f such that $g \in U_i \Longrightarrow g|V_i \sim f|V_i$ (so that in particular, g has a unique hyperbolic fixedpoint, y_i, in V_i).

Then $M \sim [V_1 \cup \ldots \cup V_k]$ is compact, and since f has no fixedpoints there, $\delta = \min\{\text{dist}(fx, x) \mid x \in M \sim \bigcup_i V_i\} > 0$. But then, of course, if the C^0-distance of g from f is less than $\delta/2$, g will also have no fixedpoints in $M \sim \bigcup V_i$. Thus, letting $U = \bigcap_{i=1}^{k} U_i \cap B^1_{\delta/2}(f)$, we have the required neighborhood of f, completing the lemma. //

In trying to prove H_n dense, we run into problems similar to those encountered in chapter 1, with the further complication that in higher dimensions, hyperbolicity is not equivalent to an easy transversality property. Our argument thus breaks into two parts. We still use transversality properties (as before) to keep control over the number of fixedpoints while perturbing, and then use further perturbations to get actual hyperbolicity. The idea of this argument is due to Jacob Palis.

We start with a temporary

Definition: $x \in \text{fix}(f)$ is a simple fixedpoint iff $(D f)_x - \text{id}$ is an isomorphism.

Simplicity is thus equivalent to 1 not being an eigenvalue of the derivative. Clearly hyperbolic fixedpoints are simple, but the latter property allows the existence of eigenvalues other than 1 (eg, -1) on the unit circle.

Lemma: All fixedpoints of f^n are simple
$\Longleftrightarrow$ $(\text{id}, f^n) \pitchfork \Delta$, where Δ = the diagonal in $M \times M$.

Proof

$(\text{id}, f^n) \pitchfork \Delta$ means that the tangent map to (id, f^n), which is (id, Tf^n), is transverse to $T\Delta$ —which is just the diagonal in $TM \times TM$. But this means that for every $x \in M$, either

$(x, f^n x) \notin \Delta$ (ie, $f^n x \neq x$) or, if $x = f^n x$, then $[(Tf^n)_x - id]$ is surjective (and hence an isomorphism). //

Since we can express simplicity as a transversality condition under a representation, we can use the Abraham transversality theorem to obtain the immediate

<u>Corollary</u>: $\widetilde{H}_n = \{f \mid fix(f^n)$ consists of only simple fixedpoints$\}$ is open and dense.

We use this corollary to help keep periodic points under control as we prove

<u>Lemma</u> (Density of H_n): For every $f \in Diff^r(M)$ and every $\varepsilon > 0$, there is $g \in H_n$ with g C^r-ε-close to f.

<u>Proof</u>

It of course suffices to show this fact assuming $f \in \widetilde{H}_n$. We note that simple fixedpoints have some useful properties of hyperbolic points: they are isolated, and they are stable; so in particular $f \in \widetilde{H}_n \Longrightarrow fix(f^n)$ is finite.

As before, we let $V_1, \ldots, V_k$ be disjoint neighborhoods of $x_1, \ldots, x_k$ — the elements of $fix(f^n)$. We can, furthermore, pick the V_i so that

$f^i V_j \cap V_k = \phi$

for $i \leq n$ and $j \neq k$.

We let $W_1, \ldots, W_k$ be smaller neighborhoods of the x_i, that is, with $\overline{W}_i \subset \overset{o}{V}_i$.

Finally, we take $\varphi_1, \ldots, \varphi_k$ to be C^k-small bump functions — that is, we want $1 \geq \varphi_i(x_i) \geq 0$, $\varphi_i = 0$ off of

W_i, and the C^k-size of φ_i small.

If we had taken V_i to be coordinate neighborhoods of the x_i (with x_i corresponding to 0) and W_i small enough that $f^n \overline{W}_i \subset \overset{o}{V}_i$, then we could define g in local coordinates in V_i as

$$g(x) = f(x) + \tilde{\varepsilon}\ \varphi_i(x) \cdot x$$

where we would pick ε a constant, smaller than ε and than $\frac{1}{3}$ the distance of the rest of the spectrum of $(D\ f)_o$ from the unit circle.

Then it is clear that g is C^r-ε-near f; we also claim: $g \in H_n$.

The latter is true because

$$(Tg^n)_{x_i} = T(f + \varepsilon\ \varphi_i \cdot id) \circ \dots \circ T(f + \varepsilon\ \varphi_i \cdot id).$$

Thus, the eigenvectors of $(Tg^n)_{x_i}$ are the same as those of $(Tf^n)_{x_i}$, but their eigenvalues have increased by $\tilde{\varepsilon}\ \varphi_i$, so that those originally on the unit circle have moved off, and no new ones have moved on; thus x_i are hyperbolic fixedpoints for g — and since f was simple, these are the only fixedpoints of g. //

Thus, we see that the study of periodic points is another area in which the generic approach proves completely satisfactory: Hartman's theorem gives us a simple description of hyperbolic points, and theorems 3 and 4 show these to be the "general case", from the generic point of view.

2.3 MORSE-SMALE SYSTEMS

The Morse-Smale systems are a generalization of the generic situation on the circle, and the case in which the theory of this chapter gives a fairly satisfactory description of orbit structure.

We begin by exploring in a little more detail the consequences of Hartman's theorem for the orbit structure of a map near a hyperbolic fixedpoint. Recall that for a linear hyperbolic map, the space is spanned by a "contracting" and an "expanding" subspace, with all other points asymptotic to these two. The conjugacy in Hartman's theorem, then, says that in a neighborhood of a hyperbolic fixedpoint, there are a "contracting submanifold" and an "expanding submanifold" running through the point, and all other points fall out of the neighborhood under both forward and backward iteration. Each of the invariant "manifolds", being a homeomorphic image of a disc in $\mathbb{R}^n$, is actually a topological manifold; it can in fact be shown (Hirsch and Pugh [1], or chap. 5 of these notes) that, if the diffeomorphism f is C^r, then these manifolds are C^r-submanifolds of M and tangent to the corresponding subspaces of TM at the point. These manifolds will be called, respectively, the <u>local stable manifold</u> and the <u>local unstable manifold</u> of the fixedpoint.

The local stable manifold of a fixedpoint x can be

characterized as the set of points that stay in a pre-assigned neighborhood of x under all forward iterates, and is denoted $W^s_{loc}(x)$. The unstable manifold can be characterized as

$$W^u_{loc}(x) = \bigcap_{m \geq 0} f^m U$$

for some neighborhood U of x. Finally we note that $W^s_{loc}(x)$ under f is the same as $W^u_{loc}(x)$ under f^{-1}, and vice-versa.

These local notions can be globalized to the <u>stable manifold</u> and <u>unstable manifold</u> of x via the definitions

$$W^s(x) = \bigcup_{n \in \mathbb{Z}^+} f^{-n}[W^s_{loc}(x)]$$

$$= \{y \mid f^n y \longrightarrow x\}$$

$$W^u(x) = \bigcup_{n \in \mathbb{Z}^+} f^{n}[W^u_{loc}(x)]$$

$$= \{y \mid f^{-n} y \longrightarrow x\} .$$

We remark that the full version of the Kupka-Smale theorem states that, in addition to hyperbolicity of all periodic points, $W^s(p) \pitchfork W^u(q)$ for p and q periodic points of a generic diffeomorphism f.

We can now define Morse-Smale systems:

<u>Definition</u>: A diffeomorphism f satisfying the following conditions is a <u>Morse-Smale diffeomorphism</u>:
(1) $\Omega(f)$ is finite (and hence $\Omega(f) = \text{per}(f)$)
(2) all periodic points are hyperbolic
(3) $W^s(p) \pitchfork W^u(q)$ for $p, q \in \Omega(f)$.

The Morse-Smale systems were initially studied by Smale ([1] and [2]); they represent the simplest kind of orbit structure that has any chance of being stable. J. Palis proved in his thesis ([2]) that Morse-Smale systems on 2- and 3-dimensional manifolds are structurally stable, and subsequently Palis and Smale ([1]) extended this result to higher dimensions.

An example of a Morse-Smale diffeomorphism is the map defined on S^2, regarded as the Riemann sphere, by

$f(z) = 2z$.

Here, the only fixedpoints are 0 and ∞ : 0 is a source, with

$W^u(0) = S^2 \sim \{\infty\}$

$W^s(0) = \{0\}$

and ∞ is a sink, with

$W^u(\infty) = \{\infty\}$

$W^s(\infty) = S^2 \sim \{0\}$.

Points "between" 0 and ∞ move away from 0 and toward ∞.

Another example, this one on T^2 (the 2-torus) is gotten by considering the time-one map of the flow given by the vectorfield $X = \text{grad } h$, where h is the height function of a slightly tilted torus (see fig. 2).

There are four fixedpoints: A, B, C, and D. A is a source, D a sink, and B and C are saddles. The curves sketched in figure 1 are the stable and unstable manifolds of the saddles, B and C. As before, $W^s(A)$ and $W^u(D)$ are each the whole torus with the other three fixedpoints deleted.

In general, if one starts with a Morse function (all critical points non-degenerate) on a compact manifold, then the flow of its gradient vectorfield has, generically, a Morse-Smale

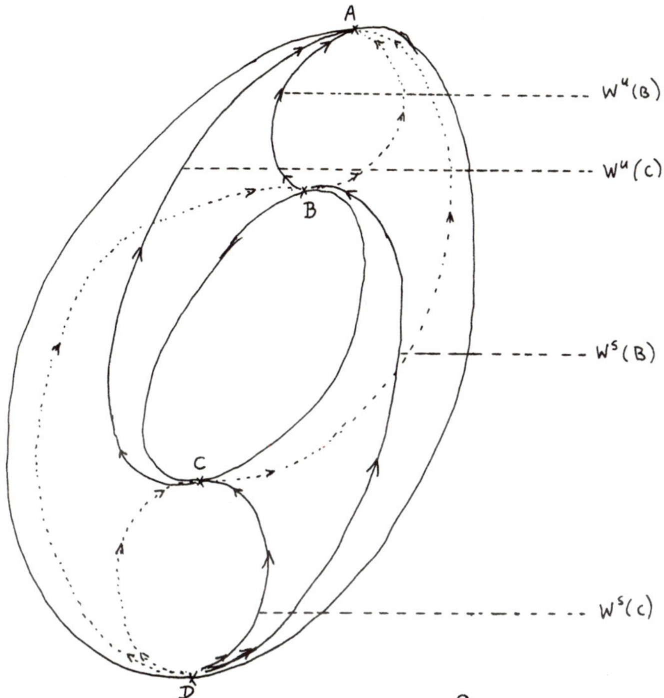

Figure 2: Morse-Smale system on T^2.

diffeomorphism as its time-one map.

This is in fact a distinguished special case of the Morse-Smale systems, called the <u>gradient-like</u> diffeomorphisms. One can define a partial ordering of the periodic points of a Morse-Smale diffeomorphism by calling $p \leq q$ if $W^u(q) \cap W^s(p) \neq \emptyset$. Then f is gradient-like if whenever $p \leq q$, $\dim W^s(p) < \dim W^s(q)$.

The phenomenon which complicates non-gradient-like diffeomorphisms is that of <u>heteroclinic points</u>. These are defined as points of (transversal) intersection of $W^u(p)$ and $W^s(q)$ where $\dim W^s(p) = \dim W^s(q)$. In this case, it is easy to see that $W^s(q)$ oscillates strongly and approaches $W^s(p)$, and likewise $W^u(q) \subset \text{clos}\, W^u(p)$.

An example on S^2 of a non-gradient-like diffeomorphism can be obtained from the gradient-like example above by modifying the map on a neighborhood, U, of 0, introducing three sinks (A, B, C) and two saddles (D, E) with stable and unstable manifolds as in figure 3.

There are some topological restrictions on the structure that can occur for the periodic points of a diffeomorphism on a given manifold.

An early example of this type of result was proven by Smale in [1]. If $\beta_i(M)$ is the $i^{\underline{th}}$ Betti number of the manifold M ($\beta_i(M) = \text{rank}\, H_i(M, \mathbb{Z})$) and $M_j(f)$ the number of periodic points of f with j-dimensional stable manifold, then for f any Morse-Smale diffeomorphism of M, the following so-called Morse inequalities hold:

$$M_o > \beta_o$$

$$M_1 - M_o \geq \beta_1 - \beta_o$$

$$\vdots$$

$$\sum_{i=o}^{\dim M} (-1)^i M_i = \sum_{i=o}^{\dim M} (-1)^i \beta_i$$

This last formula can be generaliz d for non-Morse-Smale systems by using the Lefschetz trace formula. If we consider the induced map f_{*i} on $H_i(M, \mathbb{R})$ and take its trace t_i, then we can define a homotopy invariant of f called its

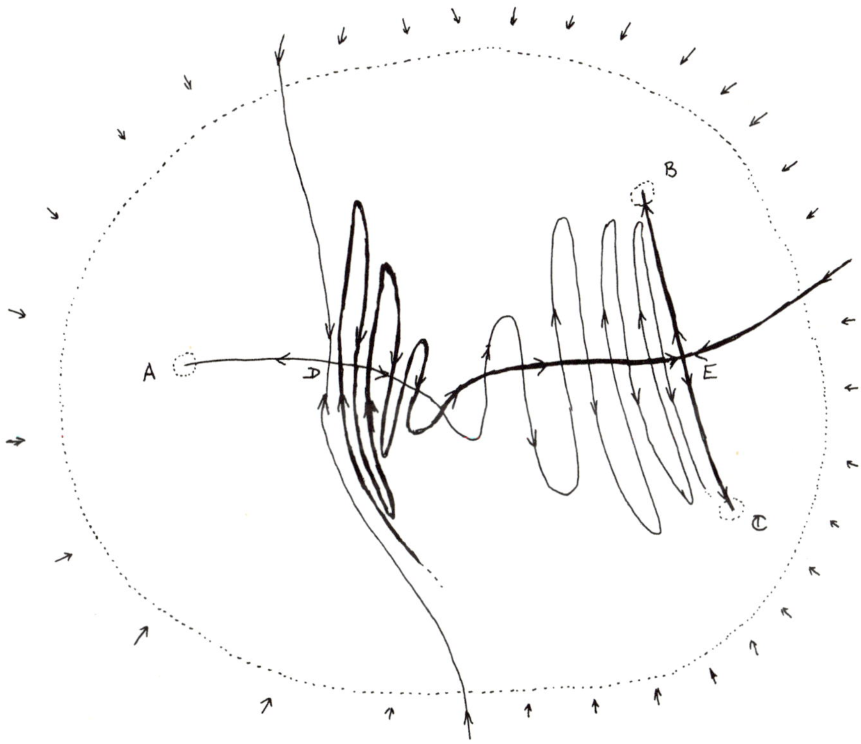

Figure 3: Modification to introduce heteroclinic points

<u>Lefschetz</u> <u>number</u>:

$$\Lambda(f) = \sum_{i=0}^{\dim M} (-1)^i \; t_i \; .$$

To each fixedpoint of f, one can attach an index, defined in local, algebraic-topology terms. In the case of hyperbolic fixedpoints, this index can be computed by the formula (Smale [3], p. 767)

$$L(p) = (-1)^u \Delta$$

where

$$u = \dim W^u(p)$$

and $\Delta = \pm 1$, respectively, when f preserves or reverses orientation on $W^u(p)$. Then the Lefschetz trace formula states that

$$\sum_{p \in \text{fix}(f)} L(p) = \Lambda(f)$$

In particular, if f is homotopic to the identity, then $\Lambda(f)$ is just the Euler characteristic of M.

CHAPTER 3

ANOSOV DIFFEOMORPHISMS

Like most fields of mathematics, the study of differentiable systems has proceeded from examples. These have served to motivate conjectures, which in turn have led either to "successful" results or to further (counter -) examples, starting the whole cycle over again with modified conjectures, and so on. From the "generic" point of view, the most important examples are those phenomena which are stable, in the sense that they occur on an open subset of $\text{Diff}^r(M)$. The obvious reason for this is the two-sided relation between generic theory and stable phenomena: that any generic property is certainly held by any stable system, and that any stable phenomenon must be accounted for in a generic description of $\text{Diff}^r(M)$.

The results so far have presented a rather blissful picture of differentiable dynamics - all phenomena up to now have been easy to describe rather completely, and Peixoto's results on the circle have even suggested that such simple systems might be generic on $\text{Diff}^r(M)$.

In this and the next chapter, we consider two "stable" classes of phenomena which present considerable complications over our picture so far; in particular, both the Anosov systems of this chapter and the "horseshoe" of the next possess an infinite number of periodic points, and hence destroy the suggestion implicit in Peixoto's results that Morse-Smale systems might be generic in higher dimensions.

The Anosov systems present a diffeomorphism analogue of Shub's expanding maps, in that (modulo certain conjectures) the non-wandering set is the whole manifold, and periodic points are dense. The first example of such a system was constructed by R. Thom as a counter-example to the conjecture that Morse-Smale systems as dense; the first published studies were by Anosov ([1], [2], [3]) arising from the context of geodesic flows on manifolds of negative curvature. Anosov systems have also been studied by ergodic theorists in the guise of "C-Systems" (Arnold and Avez [1], Sinai [1], Anosov and Sinai [1]).

Partial answers to recent conjectures about the structure of such systems have been given by Franks ([1], [2]) and Newhouse ([1]). We shall consider these and related questions at the end of this chapter.

3.1 HYPERBOLIC TORAL AUTOMORPHISMS

We begin with Thom's example on the torus, which will provide motivation for the later work.

One way of describing the (two-dimensional) torus, T^2, is to look at points on the plane, $\mathbb{R}^2$, and identify pairs of points whose coordinates differ by integers - ie, to take the quotient of the additive group $\mathbb{R}^2$ by its integer lattice. Any map f on the torus can then be "covered" (in strict analogy with the "liftings" of chapter 1) by a map which takes translates of the integer lattice into other translates of that lattice.

Conversely, we can define our example by giving a linear transformation L of $\mathbb{R}^2$. This is given by a 2×2 matrix, of which we shall require:

1) integer entries
2) determinant ± 1
3) no eigenvalue of modulus 1.

The first condition is just preservation of the integer lattice (so that L induces a well-defined map $\tilde{L}$ on T^2), the second insures that $\tilde{L}$ be a diffeomorphism, and the third is a "hyperbolicity" condition whose significance will be clarified as we proceed. For now, we note that 3) is simply the requirement that L be a hyperbolic linear map.

We begin by showing that this is an example of the type we are looking for:

<u>Lemma 1</u>: Periodic points of $\tilde{L}$ are dense in T^2.

A second proof of this fact is given in Appendix III.

<u>Proof</u>

If π denotes the projection

$$\pi\colon \mathbb{R}^2 \longrightarrow \mathbb{R}^2/_{\mathbb{Z}^2} = T^2$$

then the statement that $\pi(x)$ is periodic of period n,

$$\tilde{L}^n(\pi x) = \pi x$$

is equivalent to the statement

$$L^n(x) - x \in \mathbb{Z}^2 .$$

Thus, the set of fixedpoints of $\tilde{L}^n$ is "covered" in $\mathbb{R}^2$ by the lattice

$$(L^n - \mathrm{id})^{-1}[\mathbb{Z}^2] = \mathcal{L}_n .$$

The set of periodic points of $\tilde{L}$ is then covered by the subgroup

$$\mathcal{L}_\infty = \bigcup_{n=1}^{\infty} \mathcal{L}_n$$

of $\mathbb{R}^2$ (group property follows from the fact that if n_1 divides n_2, then $\mathcal{L}_{n_1} \subseteq \mathcal{L}_{n_2}$). Our lemma then reduces to showing that $\mathcal{L}_\infty$ is dense in $\mathbb{R}^2$.

We note that $\overline{\mathcal{L}}_\infty$ is a closed, additive subgroup of $\mathbb{R}^2$ which is invariant under all powers of L. A closed subgroup of $\mathbb{R}^2$ takes one of three forms: it is discrete, it consists of a discrete family of parallel lines, or it is $\mathbb{R}^2$.

We begin with a

Claim: $\overline{\mathcal{L}}_\infty$ contains a convergent sequence.

Note that the eigenvectors b_1, b_2 of L are also eigenvectors for all the operators of the form $(L^n - \mathrm{id})$. If the eigenvalues of L are $\lambda > 1$ and λ^{-1} (recall that $\det(L) = 1$), then the eigenvalues of $(L^n - \mathrm{id})$ are

$$\gamma_n = \lambda^n - 1$$
$$\tilde{\gamma}_n = (1 - \lambda^n)/\lambda^n \; .$$

Thus, if e_1, e_2 denote the canonical basis of $\mathbb{Z}^2$, then writing

$$e_i = \alpha_i b_1 + \beta_i b_2 \qquad i = 1, 2$$

we can generate $\mathcal{L}_n$ by

$$(L^n - \mathrm{id})^{-1} e_i = \gamma_n^{-1} \alpha_i b_1 + \tilde{\gamma}_n^{-1} \beta_i b_2 \; .$$

Note that

$$\gamma_n^{-1} \longrightarrow 0$$
$$\tilde{\gamma}_n^{-1} \longrightarrow -1$$

so that

$$\lim_{n \to \infty} (L^n - \mathrm{id})^{-1} e_i = -\beta_i b_2 \; .$$

Thus, $\overline{\mathcal{L}}_\infty$ is either $\mathbb{R}^2$ or consists of a family of parallel lines. In the latter case, $\overline{\mathcal{L}}_\infty$ contains a one-parameter subgroup, which is invariant under L. However, the only linear subspaces invariant under L are the eigenspaces. Thus, the subgroup is one of the eigenspaces.

We claim its slope is irrational. If it were rational, it would project to a circle in T^2. But an eigenspace is either contracted or expanded uniformly by L, and hence $\tilde{L}$ would either expand or contract the corresponding circle. Since a diffeomorphism can do neither, the subgroup has irrational slope.

On the other hand, a family of parallel lines, of irrational slope, which contains $\mathbb{Z}^2$, must be dense. (This last statement follows from the fact that the multiples of an irrational number, mod one, are dense in the unit interval - which is equivalent to the fact that an irrational rotation is transitive on the circle.) Thus, $\mathcal{L}_\infty$ must be all of $\mathbb{R}^2$, so that $\text{per}(\tilde{L})$ is dense in T^2 , proving the lemma . //

Thus, since periodic points are dense in T^2, the non-wandering set of $\tilde{L}$ is the whole manifold. On the other hand, the hyperbolic toral automorphisms are structurally stable. This fact was first proven by Arnold and Sinai [1], but is a special case of the Anosov stability theorem, which we shall prove in the next section.

It follows that the hyperbolic toral automorphisms are an example of an open set of diffeomorphisms whose non-wandering set is the whole manifold. We recall that although C^2 diffeomorphisms of the circle with irrational rotation number also made the whole manifold non-wandering, they were not stable.

We proceed to investigate the behavior of these examples a little more. Since the map L on $\mathbb{R}^2$ is a hyperbolic linear map, the space splits into an invariant direct sum

$$\mathbb{R}^2 = E^s \oplus E^u$$

of eigenspaces, one contracting, one expanding. We already commented in proving lemma 1 that the slope of E^s and of E^u was irrational; thus, the projection of E^s (and of E^u) is an immersed plane which winds densely about T^2. By parallel translation, we can move these two subspaces to get affine subspaces of $\mathbb{R}^2$ through each point. Their projections, which we will refer to, respectively, as $W^s(x)$ and $W^u(s)$, form two invariant foliations of T^2 by immersed planes; that is, each point belongs to a unique element of the family, and

$$\tilde{L}[W^u(x)] = W^u(\tilde{L}(x))$$
$$\tilde{L}[W^s(x)] = W^s(\tilde{L}(x))$$

If $x \in T^2$ is fixed for $\tilde{L}^n$, we note that the derivative of $\tilde{L}^n$ at x is just $L^n\colon \mathbb{R}^2 \longrightarrow \mathbb{R}^2$, so that all periodic points are hyperbolic, and $W^u(x)$, $W^s(x)$ are what we defined previously as the unstable and stable manifold of x. We can extend this definition to the whole torus by calling $W^s(x)$ the stable manifold of x, and $W^u(x)$ the unstable manifold of x, even if x is not periodic. We remark that $W^s(x)$ can, alternately, be characterized as the set of points asymptotic to x — that is, points y such that

$$\text{dist}(f^n x, f^n y) \longrightarrow 0 \quad \text{as} \quad n \longrightarrow +\infty .$$

The tangent space to T^2 at any point x is $\mathbb{R}^2$, and the (derivative) tangent map, under the canonical identifications, is just $L\colon \mathbb{R}^2 \longrightarrow \mathbb{R}^2$. Hence, the tangent space has an invariant decomposition as a direct sum

$$T[(T^2)]_x \approx \mathbb{R}^2 = E^s_x \oplus E^u_x$$

where E^s is contracted, and E^u expanded, by $(T\tilde{L})_x$.

We have already shown that periodic points of $\tilde{L}$ are dense in T^2, so that automatically, $\Omega(\tilde{L}) = T^2$. This latter fact is made somewhat plausible independently, by the fact that stable and unstable manifolds of a fixedpoint wrap densely around the torus. It becomes even more plausible in view of

Lemma 2: λ, the eigenvalue of L, is irrational.

(Recall that λ is the rate of "expansion" along E^u, and of "contraction" along E^s, and that "irrational" motion on a compact manifold tends to be dense.)

Proof

The characteristic polynomial of L can be written two ways:

$$P(x) = (x - \lambda)(x - \frac{1}{\lambda})$$

$$P(x) = x^2 - (\mathrm{tr}\, L)x + (\det L) .$$

But the characteristic polynomial is an integer polynomial (since L is an integer matrix), and hence the trace of L is an integer.

Now, assume λ is rational. Write $\lambda = \frac{p}{q}$ in lowest terms. Then

$$\mathrm{tr}\, L = \lambda + \frac{1}{\lambda} = \frac{p}{q} + \frac{q}{p} = \frac{p^2 + q^2}{pq} .$$

But since $\mathrm{tr}\, L \in \mathbb{Z}$, this means $p^2 + q^2$ is divisible by pq, hence by p. Since p clearly divides p^2, this means p also divides q^2, contradicting the assumption that λ was written in lowest terms. //

3.2 THE ANOSOV STABILITY THEOREM

We now proceed to generalize the toral examples above to the class of Anosov diffeomorphisms. The property that becomes central is the "hyperbolicity" condition referred to rather cryptically in the last section. Recall that on the torus, we have a canonical identification of the tangent spaces at a point and at its image, relative to which the derivative of our toral examples is hyperbolic —— so that we have (by another characterization of hyperbolicity) an invariant decomposition of the tangent space into eigenspaces. In general, while we have no canonical identification of tangent spaces (ie, the tangent bundle is not trivial), we can generalize the eigenspace decomposition as follows:

Definition: $f \in \mathrm{Diff}(M)$ is an Anosov diffeomorphism if there exists a splitting of the tangent bundle to M into a (continuous) Whitney sum of sub-bundles

$$TM = E^s \oplus E^u$$

such that Tf preserves the splitting and, relative to some Riemann metric $\| \cdot \|$ on TM, there exist real constants $c > 0$ and $\lambda > 1$ such that

$$\| Tf^n v \| \leq c \lambda^{-n} \| v \| \quad \text{for } v \in E^s$$
$$\| Tf^n v \| \geq c \lambda^{n} \| v \| \quad \text{for } v \in E^u$$

for every $n \in \mathbf{Z}^+$.

We remark that if M is compact, then f satisfies or violates the above conditions (possibly with varying $c > 0$ and $\lambda > 1$) in all Riemann metrics simultaneously. Furthermore, in this compact case, a process similar to that in the lemma on p. 71, chapter 2 (due to Mather) allows us to renorm the manifold so that $c = 1$.

There is an alternate characterization of Anosov diffeomorphisms in terms of functional analysis, which we shall use in proving theorems. As before, we let $\Gamma_b = \Gamma_b(TM)$ denote the Banach space of bounded vectorfields on M. Let $f \in \text{Diff}(M)$.

Lemma 3: Let $f^*: \Gamma_b \longrightarrow \Gamma_b$ be the linear operator defined by

$$(f^* \gamma)_x = Tf(\gamma_{f^{-1}x}) .$$

Then f is Anosov iff f^* is hyperbolic.

Note that this loosely corresponds to the hyperbolicity of L in the last section. In the compact case, the above condition is equivalent to saying that id_M is a hyperbolic fixedpoint of conjugation by f as an operator on the Banach manifold of bounded maps on M.

Proof

$\Longrightarrow$: If f is Anosov, this is easy: define Γ_b^s to be the space of all (bounded) vectorfields mapping into E^s, and Γ_b^u those with image in E^u. It follows almost immediately from the definitions that f^* contracts Γ_b^s and expands Γ_b^u, so that f^* is hyperbolic.

$\Longleftarrow$: The difficulty in proving the converse is that, while we are given a priori a splitting of Γ_b under f^*,

$$\Gamma_b = \Gamma_b^s \oplus \Gamma_b^u ,$$

we cannot be sure that the union of images of elements of Γ_b^s and of Γ_b^u will give a splitting of TM. They will clearly fill everything, but they might overlap.

We solve the problem by a trick. Given $v \in TM$, we wish to decompose it into a "contracting" and an "expanding" part. We lift the problem to Γ_b, decompose there, and evaluate to get back to TM; then we show that this decomposition downstairs is the only possible one.

To lift v, define the bounded vectorfield

$$\sigma_v(x) = \begin{cases} v & \text{if } v \in TM_x \\ 0 & \text{otherwise} \end{cases}.$$

Thus, σ_v takes the value v where it can, and vanishs elsewhere.

Since f^* is hyperbolic, we can decompose σ_v uniquely:

$$\sigma_v = \gamma_s \oplus \gamma_u .$$

If we evaluate to get

$$v_s = \gamma_s(x), \quad v_u = \gamma_u(x),$$

then we will be done if we can show

$$\sigma_{v_s} \in \Gamma_b^s$$

$$\sigma_{v_u} \in \Gamma_b^u$$

where σ_{v_s}, σ_{v_u} are defined analogously with σ_v; for clearly

$$\sigma_v = \sigma_{v_s} + \sigma_{v_u}$$

and so by uniqueness of the splitting

$$\Gamma_b = \Gamma_b^s \oplus \Gamma_b^u$$

we would have

$$\gamma_s = \sigma_{v_s}$$
$$\gamma_u = \sigma_{v_u}$$

which then clearly implies that $TM = \operatorname{im} \Gamma^s \oplus \operatorname{im} \Gamma^u$, and that Tf "contracts" Γ_b^s and "expands" Γ_b^u.

We show $\sigma_{v_s} \in \Gamma_b^s$:

$$\| (f^*)^n(\sigma_{v_s}) \|_{(\Gamma_b)}$$
$$= \| Tf^n \circ \sigma_{v_s} \circ f^{-n} \|_{(\Gamma_b)}$$

but $\sigma_{v_s} = 0$ except at x ($v \in TM_x$) so that

$$\| Tf^n \circ \sigma_{v_s} \circ f^{-n} \|_{(\Gamma_b)} = \sup_{z \in M} \| Tf^n \sigma_{v_s}(f^{-n}z) \|_{(TM)}$$
$$= \| Tf^n \sigma_{v_s}(z) \|_{TM}$$
$$= \| Tf^n v_s \|_{TM} = \| Tf^n \gamma_s(x) \|_{TM}$$
$$\leq \sup_{z \in M} \| Tf^n \gamma_s(f^{-n}_z) \|_{TM}$$
$$= \| (f^*)^n \gamma_s \|_{\Gamma_b}$$
$$\leq \lambda^{-n} \| \gamma_s \|_{\Gamma_b} \longrightarrow 0 \quad \text{as} \quad n \longrightarrow +\infty$$

But by definition, $\gamma \in \Gamma_b^s$ iff

$$\| (f^*)^n \|_{\Gamma_b} \longrightarrow 0 \quad \text{as} \quad n \longrightarrow \infty$$

hence, $\sigma_{v_s} \in \Gamma_b^s$.

Similarly, $\sigma_{v_u} \in \Gamma_b^u$, and we are done. //

<u>Note</u>: (1) We used bounded instead of continuous vectorfields to be able to define σ_v. It is possible, however, to define

a continuous version of σ_v and, using somewhat trickier arguments, to show the same result with Γ_0 (continuous vectorfields) instead of Γ_b. (2) In the compact case, Mather has refined this characterization in [1]: one need only assume f^* is simple (1 is not an eigenvalue) to prove f Anosov.

In Appendix IV, we sketch a proof of the following

Proposition: The collection of Anosov diffeomorphisms is open in $\mathrm{Diff}^r(M)$ $(r \geq 1)$.

That is, a C^r-perturbation of an Anosov diffeomorphism is Anosov. We want now to prove a different perturbation result, that Anosov diffeomorphisms are structurally stable. Our proof will exploit another property of Anosov diffeomorphisms, of some independent interest.

Definition: A map f of a metric space X to itself is expansive iff there exists a uniform constant $\varepsilon > 0$ such that for any pair of points of X, $x \neq y$, there is an iterate (possibly negative, when f is invertible) such that

$\mathrm{dist}\,(f^n x, f^n y) \geq \varepsilon$.

The most obvious example of expansive maps is the expanding maps of chapter 1. Another is provided by

Lemma 4: Hyperbolic linear isomorphisms are expansive.

Proof

Given $x \neq y$, decompose each point into its "contracting" and "expanding" parts: $x = x_s \oplus x_u$, $y = y_s \oplus y_u$. Since $x \neq y$, either $x_s \neq y_s$ or $x_u \neq y_u$. In the latter case, as $n \longrightarrow +\infty$,

$$\| L^n(x_u - y_u) \| \longrightarrow \infty, \quad \| L^n(x_s - y_s) \| \longrightarrow 0$$

so that

$$\| L^n x - L^n y \| = \| L^n(x - y) \| \longrightarrow \infty .$$

In the former case, as $n \longrightarrow +\infty$,

$$\| L_u^{-n}(x_u - y_u) \| \longrightarrow 0, \quad \| L^{-n}(x_s - y_s) \| \longrightarrow \infty ,$$

so that

$$\| L^{-n}x - L^{-n}y \| = \| L^{-n}(x-y) \| \longrightarrow \infty$$

and L is expansive. //

A third example, and the reason for our interest in this property, is

Lemma 5: Anosov diffeomorphisms of compact manifolds are expansive.

Proof

We exploit lemmas 3 and 4. Consider the diffeomorphism F defined on a neighborhood of the zero-section in $\Gamma_b(TM)$ by

$\exp[F(v)] = f \circ \exp \circ v \circ f^{-1}$.

Then the zero-section is a fixedpoint of F, with derivative f^*. Thus, by Hartman's theorem (chapter 2), we have a local conjugacy between f^* and F. In particular, for a neighborhood on which the conjugacy is defined, any nonzero vectorfield σ has some iterate under F outside the neighborhood.

Let ε denote the diameter of this neighborhood. With a little care, we can make sure that $\exp_x$ maps diffeomorphically onto the ε-ball about x, for every $x \in M$. Given $y \neq x$ in

M, either y is outside the ε-ball about x to begin with, or we can define a bounded vectorfield so that $\sigma(x) = \exp_x^{-1} y$ and $\sigma(z) = 0$ for $z \neq x$. In this latter case, some iterate of σ under F lies outside the ε-ball about 0. But $(F\sigma^n)(f^n x) = f^n y$ and $(F\sigma^n)(z) = 0$ for $z \neq f^n x$.

Hence, $f^n y$ lies outside the ε-ball about $f^n x$, so that f is expansive. //

We are now ready to show

Theorem 1 (Anosov Stability Theorem): Anosov diffeomorphisms on compact manifolds are structurally stable.

Before giving a proof, we briefly outline the history of this theorem. The stability if hyperbolic toral automorphisms was announced by Arnold and Sinai in [1] with an incorrect proof; the first general studies of Anosov diffeomorphisms were the papers [1] and [2] of Anosov. He proved the general stability theorem ([1]) and ([3]) using an appropriate generalization of the stable and unstable foliations of 3.1 and a direct geometric argument. A function-theoretic proof is given by Moser in [1]; the proof we give is essentially a "global" reformulation of this proof. An alternate "global" version of Moser's proof is given by Mather in an appendix to Smale's survey [3], but this version contains certain technical difficulties. Recently, Shub has given a third proof (unpublished), based on his "equivariant fibration theorem" (Hirsch, Pugh, and Shub [1]).

Our version exploits the stability of hyperbolic fixedpoints (theorem 3, chapter 2).

<u>Proof of Theorem 1</u>

We put the problem into a function-analytic framework by defining the operator $\mathcal{H}_g$ on $C^o(M)$ via

$$\mathcal{H}_g = \omega_g \circ \alpha_{f^{-1}}$$

or equivalently,

$$\mathcal{H}_g(h) = g \circ h \circ f^{-1}$$

where f is our given Anosov diffeomorphism and g will be a diffeomorphism C^1 near f. It is an immediate consequence of the theory in chapter 0 that $\mathcal{H}_g$ is a C^1 map (for g C^1) and varies continuously in the C^1 topology as g varies in $C^1(M)$. We

<u>Claim</u>: $\mathcal{H}_f$ has id_M has a hyperbolic fixedpoint.

This is clear from computing the derivative of $\mathcal{H}_f$ at id_M:

$$(D\mathcal{H}_f)_o = f^* .$$

Thus, by the stability of hyperbolic fixedpoints, $\mathcal{H}_g$ will have a fixedpoint h C^o-near id_M, for g C^1-near f. That is, h will be a continuous map satisfying $h = g \circ h \circ f^{-1}$, or $h \circ f = g \circ h$, or equivalently, making

$$\begin{array}{ccc} M & \xrightarrow{f} & M \\ {\scriptstyle h}\downarrow & & \downarrow{\scriptstyle h} \\ M & \xrightarrow[g]{} & M \end{array}$$

commute. That is, we have a semi-conjugacy. To make g

conjugate to f, we need to prove h a homeomorphism. We begin with the

Claim: h is injective, provided the C^0 distance of h from id_M is less than $\frac{1}{2}\varepsilon$, where ε is the expansive constant of f.

Suppose $x \neq y$. Then for some n,

$$\begin{aligned}\varepsilon &< d(f^n x, f^n y) \\ &\leq d(f^n x, h f^n x) + d(hf^n x, hf^n y) + d(hf^n y, f^n y) \\ &\leq \frac{\varepsilon}{2} + d(hf^n x, hf^n y) + \frac{\varepsilon}{2} \\ &= \varepsilon + d(g^n hx, g^n hy) .\end{aligned}$$

Hence, $d(g^n hx, g^n hy) > 0$, so clearly $hx \neq hy$, and h is injective.

We point out that C^0 distance of h from id_M can be made small by making g C^1-near f.

Now, to show h a homeomorphism, we note that h, being a continuous injection of a compact space into a Hausdorff space, is a homeomorphism with its image. We then use invariance of domain to show the image of h open and hence (since also compact) the whole manifold, and thus h is a homeomorphism, completing the proof of the theorem. //

3.3 GENERAL QUESTIONS

The definition of Anosov diffeomorphisms given in the last section is, of course, simply a translation into a more general setting of some properties of the hyperbolic toral automorphisms. The question naturally arises whether the more general definition represents a real extension of the toral examples. We begin this section by giving the most general known construction of Anosov diffeomorphisms, which in particular gives non-toral examples, one of which we shall compute. The technique is basically given by Smale in [3] and extended by Shub in [1] to the general form here.

Suppose we are given a Lie group G, expressible as a semi-direct product $G = N \cdot C$, where N is a real connected, simply connected nilpotent Lie group with a left-invariant metric, and C a finite group of automorphisms of N. Let Γ be a discrete uniform subgroup of G —— ie, a discrete subgroup whose coset space N/Γ is compact.

By a theorem of L. Auslander ([1]), we can assert

(1) $\Gamma \cap N$ is a uniform, discrete subgroup of N
(2) $\Gamma \cap N$ has finite index in Γ
(3) $N/_{\Gamma \cap N}$ is a finite covering space of N/Γ.

Now, suppose we are given an automorphism H of N which is hyperbolic in the sense that its derivative at the origin is hyperbolic (or equivalently, that the induced Lie algebra automorphism has no eigenvalues on the unit circle). Suppose furthermore that H preserves Γ in the following sense: for every $\gamma \in \Gamma$, there is a $\gamma_H \in \Gamma$ such that $H(\gamma \cdot g) = \gamma_H \cdot H(g)$ for every $g \in N$; or equivalently, $H \Gamma H^{-1} \subset \Gamma$ as subgroups of $\mathrm{Diff}(N)$.

Then the map on cosets defined by

$$\Gamma \cdot x \longrightarrow \Gamma \cdot H(x)$$

defines an Anosov map

$$H_o : N/\Gamma \longrightarrow N/\Gamma$$

which is a diffeomorphism provided $H \Gamma H^{-1} = \Gamma$, and is finitely covered by a map

$$H_1 : N/\Gamma \cap N \longrightarrow N/\Gamma \cap N$$

which is an Anosov diffeomorphism of the nilmanifold $N/\Gamma \cap N$. In case $C = \{1\}$, of course $G = N$, $\Gamma = \Gamma \cap N$, $H_o = H_1$; in either case, H_1 is called a <u>hyperbolic nilmanifold automorphism</u> and H_o is called a <u>hyperbolic infranilmanifold automorphism</u>. (The latter terminology is due to Hirsch: N/Γ is finitely covered by the nilmanifold $N/\Gamma \cap N$, and is therefore an "infra"-nilmanifold.

We compute two examples, given in Smale [3] and Shub [1], which in particular show the existence of non-toral Anosov diffeomorphisms.

We begin by taking G_1 and G_2 to be two copies of the (unique non-trivial) 3-dimensional real nilpotent Lie group. G_i can be thought of as given by 3×3 matrices of the form

$$\begin{pmatrix} 1 & 0 & 0 \\ x & 1 & 0 \\ z & y & 1 \end{pmatrix}$$

with x, y, z arbitrary real numbers. We let $N = G_1 \times G_2$, or equivalently, 6×6 matrices of the form

$$\begin{pmatrix} g_1 & 0 \\ 0 & g_2 \end{pmatrix}$$

with $g_i \in G_i$. Thus, an element of N is determined by the six entries below the diagonal in g_1 and g_2; we will (for clarity) suppress all other entries, which are canonically 0's and 1's. We let $\Gamma_0 = \Gamma \cap N$ be the subgroup of matrices in N with entries in $\mathbb{Z}(\sqrt{3})$ —— ie, of the form $\alpha + \beta\sqrt{3}$, with α and β integers. C will be the group of automorphisms of N generated by the identity and the automorphism

$$\begin{pmatrix} a & & & \\ b & c & & \\ & & d & \\ & & e & f \end{pmatrix} \longmapsto \begin{pmatrix} -a & & & \\ -b & c & & \\ & & -d & \\ & & -e & f \end{pmatrix}$$

which we denote by B. Clearly, $B^2 = \mathrm{id}$, so C is a group with two elements. Now, we need to pick constants of "expansion" and "contraction" for a candidate for H above. Let $\lambda = 2 + \sqrt{3}$. We define a "conjugation" operator on $\mathbb{Q}(\sqrt{3})$ to be the automorphism $x \longmapsto \bar{x}$ generated over $\mathbb{Q}$ by $\overline{\sqrt{3}} = -\sqrt{3}$. Pick $w \in \mathbb{Q}(\sqrt{3})$ such that $w \notin \mathbb{Z}(\sqrt{3})$, but $2w \in \mathbb{Z}(\sqrt{3})$ and $(\lambda-1)w \in \mathbb{Z}(\sqrt{3})$. Thus, w is of the form $w = \frac{m}{2} + \frac{n}{2}\sqrt{3}$, with m and n odd. We can then define B_1 to be the element of N given by

$$B_1 = \begin{pmatrix} 0 & & & \\ 0 & w & & \\ & & 0 & \\ & & 0 & \bar{w} \end{pmatrix}.$$

B_1 acts on N by inner automorphisms, and we can let Γ be the subgroup of $N \cdot C$ generated by Γ_o and B_1; so then $\Gamma = \Gamma_o \cup B_1 \Gamma_o$. Finally, we define two automorphisms of N as follows

$$H \begin{pmatrix} a & & & \\ b & c & & \\ & & d & \\ & & e & f \end{pmatrix} = \begin{pmatrix} \lambda a & & & \\ \lambda^2 b & \lambda^3 c & & \\ & & \lambda^{-1} d & \\ & & \lambda^{-2} e & \lambda^{-3} f \end{pmatrix}$$

$$\tilde{H} \begin{pmatrix} a & & & \\ b & c & & \\ & & d & \\ & & e & f \end{pmatrix} = \begin{pmatrix} \lambda a & & & \\ \lambda^{-3} b & \lambda^{-2} c & & \\ & & \lambda^{-1} d & \\ & & \lambda^{3} e & \lambda^{2} f \end{pmatrix}.$$

Then $H \Gamma H^{-1} = \Gamma = \tilde{H} \Gamma \tilde{H}^{-1}$, so that we have induced diffeomorphisms H_1, $\tilde{H}_1$ of the nilmanifold N/Γ_o and H_o, $\tilde{H}_o$ of the infranilmanifold N/Γ; these are all clearly Anosov. We note that although the splitting in $G = N \cdot C$ gives a contracting and expanding subgroup, in the second case, G is not (group-theoretically) the product of these subgroups. It is also reasonably clear that N/Γ is neither a torus nor even a nilmanifold, so that these examples give non-trivial generalizations of those of 3.1.

The process described above is the most general known method of constructing Anosov diffeomorphisms. It is natural to ask if this is essentially the most general construction possible. This was formulated by Smale as a

<u>Conjecture</u>: Every Anosov diffeomorphism on a compact manifold is topologically conjugate to a hyperbolic infranilmanifold automorphism.

Note that this conjecture says that any compact manifold supporting an Anosov diffeomorphism is (up to change if differentiable structure) finitely covered by a compact nilmanifold, and that the map has (up to conjugacy) a structure strongly related to the homogeneous structure of this finite covering.

The most general results concerning this conjecture are given by Franks in [1]. There he proves a strong version of the conjecture under two assumptions:

(1) $\Omega(f) = M$

(2) Either E^s or E^u is one-dimensional

Franks has also shown ([2]) that if the underlying manifold is a torus to begin with, then the map is conjugate to a hyperbolic automorphism, given (2) and certain assumptions on the induced homology map.

A second conjecture concerns assumption (1) above. It follows from the Anosov Stability Theorem and the General Density Theorem that for an Anosov diffeomorphism, periodic points are dense in the non-wandering set. This can also be shown directly (see Franks [1], p. 14 or Anosov [3], p. 116; there is also a cute proof in Walters [1], p. 79.) Thus, assumption (1) is equivalent to the obvious generalization of Lemma 1 above:

<u>Conjecture</u>: The periodic points of an Anosov diffeomorphism of a compact manifold M are dense in M.

This conjecture is also equivalent to the existence of an invariant Lebesgue measure for f. Anosov showed ([2]) that the existence of a measure implies density of periodic points (and in fact, ergodicity of f), while Sinai ([1]) and Bowen ([1], in a more general case) have constructed an invariant, ergodic measure for f, assuming periodic points dense. The only positive result in the direction of this conjecture is due to Newhouse ([1]), who shows that assumption (2) above implies (1).

Finally, Walters ([1]) shows another interesting stability property of Anosov diffeomorphisms —— topological stability: a C^0-perturbation g of Anosov diffeomorphism f is semi-conjugate to it via a continuous surjection making the diagram

$$\begin{array}{ccc} M & \xrightarrow{g} & M \\ h\downarrow & & \downarrow h \\ M & \xrightarrow[f]{} & M \end{array}$$

commute.

CHAPTER 4
THE HORSESHOE

In this chapter, we study an example originally developed by Smale ([4]) to show that Morse-Smale systems are not dense in $\mathrm{Diff}^r(M)$. The example is in some sense even more complicated than the Anosov systems of chapter 3; while there are an infinite number of periodic points, the non-wandering set is neither the whole manifold nor a submanifold - in fact, it will turn out to contain an isolated "piece" which is closed, perfect, and totally disconnected, - in other words, a "cantor" set. The example is stable, even in the strong sense that it can be embedded in a structurally stable system. We shall prove directly (and in very bare-handed fashion) that the example is Ω-stable. This fact follows from the Ω-stability theorem, which we shall discuss in chapter 6, but the direct proof for this case is instructive since it serves (as it did historically) to motivate most of the ideas in the more general result. On the other hand, the structural stability follows from a recent general result of Robbin ([1]).

The chapter treats three topics. We need a "computational" model for the behavior of our example; this is provided by the shift automorphism of symbolic dynamics, which we present first. Second, we build the example and show how it can be modelled on the shift automorphism; in the process we show that perturbations can also be modelled on the shift automorphism, in similar fashion, thus proving Ω-stability. Finally, we show that this phenomenon exists in mathematical (and physical!) "nature", in the guise of homoclinic points.

As far as I know, there are two written accounts of the horseshoe: Smale's original exposition ([4]) and a set of unpublished notes by Sacker ([1]). Both follow the pattern above, and differ relatively little; the presentation here is a digested synthesis of the other two. The notation is essentially the same in both papers, and we continue the tradition.

4.1 THE SHIFT AUTOMORPHISM

The most wide-ranging references on this subject are Gottschalk and Hedlund ([1]) and Morse ([1]); however both Smale ([4]) and Sacker ([1]) treat the relevant parts of this theory.

Briefly, the shift automorphism can be described as follows: take the collection of all two-sided sequences ("bisequences")

$$\cdots a_{-1} a_0 a_1 \cdots$$

of positive integers less than or equal to n, topologized by taking as a base set the collection of all bisequences having specified entries in a finite number of specified positions; then the shift automorphism acts on a bisequence by "shifting" all of its entries one place to the left.

For technical reasons, however, we shall be a bit more formal in our definitions. We start with a set $\mathcal{S}_n$ of n <u>symbols</u>. We can for example take $\mathcal{S}_n$ to be the set

$\mathcal{S}_n = \{0, 1, \ldots, n-1\}$.

Given two integers ℓ, j with $-\ell \leq j$, we define A_j to be the set of finite "sequences"

$a_{-\ell}, a_{-\ell+1}, \ldots, a_j$

of symbols from $\mathcal{S}_n$; this is of course the same as calling A_j the space $\mathcal{S}^{[-\ell, j]}$ of maps from $[-\ell, j]$ (by which, in this section, we shall mean the set of integers between $-\ell$ and j) to $\mathcal{S}$.

We then have three alternate descriptions of A_∞^∞ :

(1) First, of course, a natural extension of the definition of A_j says that A_∞^∞ should be the collection of all bisequences in $\mathcal{S}$, with the topology defined by the subbase of sets $B(\ell, j, \{a_n\}) = \{\{b_n\} | b_{-\ell} = a_{-\ell}, \ldots, b_j = a_j\}$.

(2) Second, we can give each of the A_j the discrete topology and note that for $\ell_1 \leq \ell$, $j_1 \leq j$, we have a natural "restriction" map

$p: A_j^\ell \longrightarrow A_{j_1}^{\ell_1}$

gotten by chopping off the extra terms. Then the p's form an inverse limit system, whose inverse limit is (set-theoretically) A_∞^∞. We can describe the topology on A_∞^∞ as the inverse limit if the discrete topologies on A_j^ℓ.

(3) Finally, a more succinct form of the above is to call A_∞^∞ the weak direct product of a bisequence of copies of the (discrete) topological space S :

$$A_\infty^\infty = \prod_{K \in \mathbb{Z}} S .$$

These descriptions are equivalent, and the topology on A_∞^∞ can be meterized; two metrics occur naturally: one is

$$d(\{a_n\}, \{b_n\}) = \frac{1}{1+N}$$

where

$$N = \sup\{n \mid a_i = b_i \text{ for all } i \text{ with } |i| \leq n\}$$

the other is

$$d(\{a_n\}, \{b_n\}) = \sum_{n=-\infty}^{\infty} \frac{1}{2^{|n|}} \delta_n ,$$

where $\delta_n = \begin{cases} 0 & \text{if } a_n = b_n \\ 1 & \text{if } a_n \neq b_n . \end{cases}$

By the Tychynoff theorem and description (3), A_∞^∞ is compact. It is fairly easy to see that A_∞^∞ is perfect and totally disconnected; being a compact metric space, it is therefore a Cantor discontinuum. In particular, (see Hocking and Young [1], p. 97, for a proof) it is homeomorphic to the cantor middle-third set. A more natural embedding in $\mathbb{R}$, however, is to assign to a sequence of symbols chosen from $\{0, \ldots, n-1\}$ the number whose $(n+1)$-ary expansion is just that sequence. The image of this embedding is then the set of points with no "n" in their expansion. In the case when $n = 2$, if instead of mapping the symbol "1" to 1 we map it to 2, then the intersection of the embedded set with the unit interval will be the usual Cantor middle-third set.

We can use description (2) to define the shift automorphism as follows: for each pair (ℓ, j) (for which A_j^ℓ exists), there is a well-defined bijection

$$\tau_j^\ell : A_j^\ell \longrightarrow A_{j-1}^{\ell+1}$$

by simply renaming places - thus, the entries of an element of A_j^ℓ are indexed from $-\ell$ to j, and we make it an element of $A_{j-1}^{\ell+1}$ by re-indexing, starting at $-(\ell+1) = -\ell-1$. Clearly the diagram

$$\begin{array}{ccc} A_j^\ell & \xrightarrow{\tau_j^\ell} & A_{j-1}^{\ell+1} \\ {\scriptstyle p}\downarrow & & \downarrow{\scriptstyle p} \\ A_{j'}^{\ell'} & \xrightarrow[\tau_{j'}^{\ell'}]{} & A_{j'-1}^{\ell'+1} \end{array}$$

commutes whenever $\ell' \leq \ell$, $j' \leq j$, so that the τ_j^ℓ lift to a map on the inverse limit, which we call τ :

$$\tau : A_\infty^\infty \longrightarrow A_\infty^\infty .$$

When A_∞^∞ comes from $\mathcal{S}_n$, we call τ the <u>shift automorphism on n symbols.</u> Since each τ_j^ℓ was a homeomorphism, so is Under the $(n+1)$-ary representation described above τ is represented by multiplication by $(n+1)$.

We wish to say something about the periodic points of τ. Note that

$$\tau^n \{\cdots a_{-1}\, a_0\, a_1 \cdots\} = \{\cdots a_{n-1}, a_n, a_{n+1} \cdots\}$$

so that a bisequence

$\cdots x_{-1}\, x_o\, x_1 \cdots$

is periodic of period n if and only if

$$\begin{aligned} x_{-1} &= x_{n-1} \\ x_o &= x_n \\ x_1 &= x_{n+1} \\ &\vdots \\ x_k &= x_{n+k} \end{aligned}$$

or in other words, the sequence is just the reiteration of a single block of n symbols.

We <u>claim</u>: periodic points of τ are dense in A_∞^∞. To show this, we note that one bisequence is close to another iff they agree on some finite block about the zero-slot. Hence, we define a sequence of periodic bisequences

$p_k = \{\cdots a_{-1}^{(k)}\; a_o^{(k)}\; a_1^{(k)}\; \cdots\}$ converging to a given sequence

$p_o = \{\cdots a_{-1}^{(o)}\; a_o^{(o)}\; a_1^{(o)}\; \cdots\}$

by iterating the first k-block of p_o :

$a_{i+jk}^{(k)} = a_i^{(k)} = a_i^{(o)} \qquad (|i| < k,\;\; j \in \mathbb{Z}^+)$

Since periodic points are non-wandering we see also that p_o is non-wandering for τ . We summarize the important properties of τ in

<u>Lemma 1</u>: (1) A_∞^∞ is a cantor set
(2) τ is a homeomorphism
(3) $\Omega(\tau) = A_\infty^\infty$, and periodic points are dense in A_∞^∞ .

4.2 DEFINITION OF THE HORSESHOE

We begin by giving a very sloppy "geometric" picture (via a

homotopy from the identity) of the horseshoe, before we proceed to establish notation and give a more formal definition.

Start with a rectangle, R, in the plane, as the domain of the map. Squeeze this rectangle along one direction (the horizontal) and stretch it along the other (vertical), until we have a long, thin "bar". Then bend this bar somewhere near its middle, to get a "horseshoe" shape, which we then put back "over" the original rectangle in such a way that the bent part and the ends of the original "bar" project outside the rectangle:

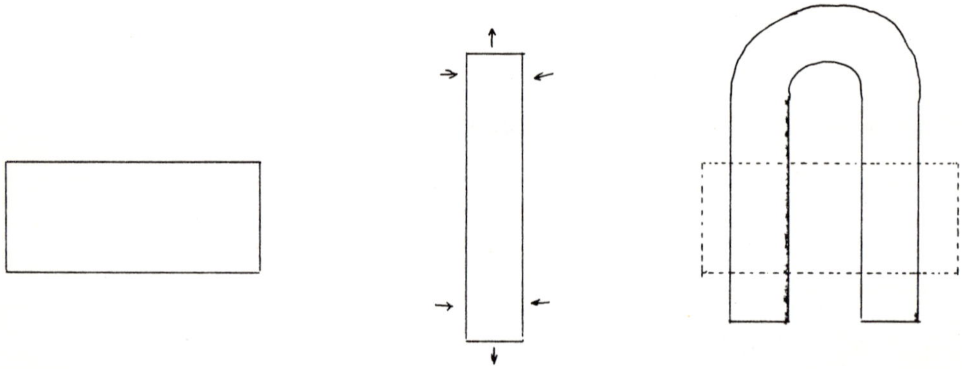

(a) rectangle (b) "bar" (c) "horseshoe"

We can invoke a theorem of Palais ([4]) to show that this map (which can clearly be made smooth) can be extended to a diffeomorphism of some disc containing the rectangle and the image, so that we will be able in particular to embed this map in any surface.

To define the map more formally and more precisely, we start by writing the rectangle R as a product of two closed intervals, I and J, embedded in the plane, and with the projections

$\pi_1 : R = I \times J \longrightarrow I$

$\pi_2 : R = I \times J \longrightarrow J$.

We pick two pairs of disjoint closed intervals, $I_1, I_2 \subset \overset{\circ}{I}$ and $J_1, J_2 \subset \overset{\circ}{J}$ ("o" = "interior"), and define $R_1, R_2 \subset R$ by $R_i = \pi_1^{-1}[I_i]$ $(i = 1,2)$

Our picture so far is

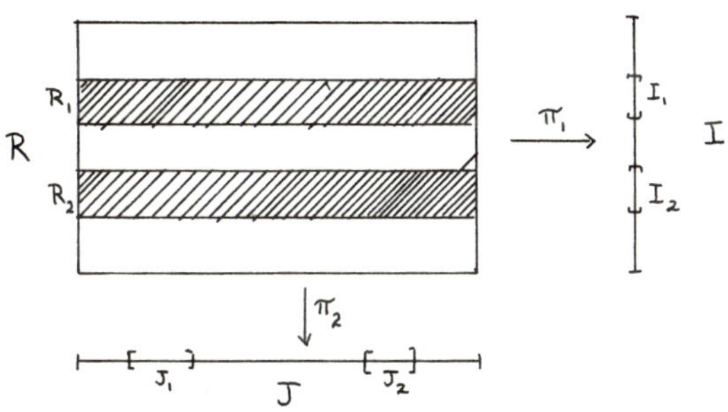

We now define the diffeomorphism (into), $f: R \longrightarrow \mathbb{R}^2$, by requiring

(1) $f[R_i] = \pi_2^{-1}[J_i]$ $(i = 1,2)$

(2) $f|R_i$ is affine, and expressible by a matrix of the form

$$\begin{pmatrix} \lambda_1^{(i)} & 0 \\ 0 & \lambda_2^{(i)} \end{pmatrix} \quad \text{with} \quad \begin{array}{l} (-1)^i \lambda_1^{(i)} > 1 \\ 0 < (-1)^i \lambda_2^{(i)} < 1 \end{array}$$

(3) $f[R \sim (R_1 \cup R_2)] \subseteq \mathbb{R}^2 \sim R$ (we can assume further that $f[R]$ is contained in some neighborhood of R — for example, $2I \times J$) .

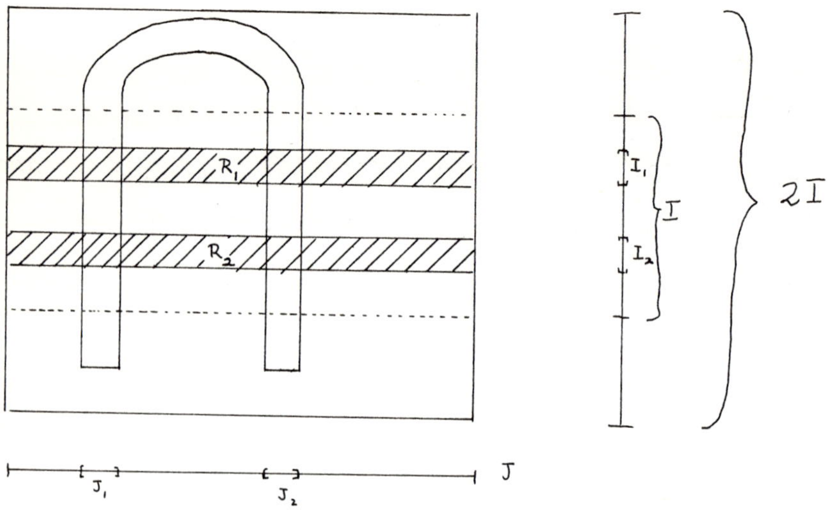

We can extend such a map on R to a map of $2I \times J$ into itself by putting little "caps" on the ends of the horseshoe, and contracting the extra upper and lower parts of $2I \times J$ into the respective caps:

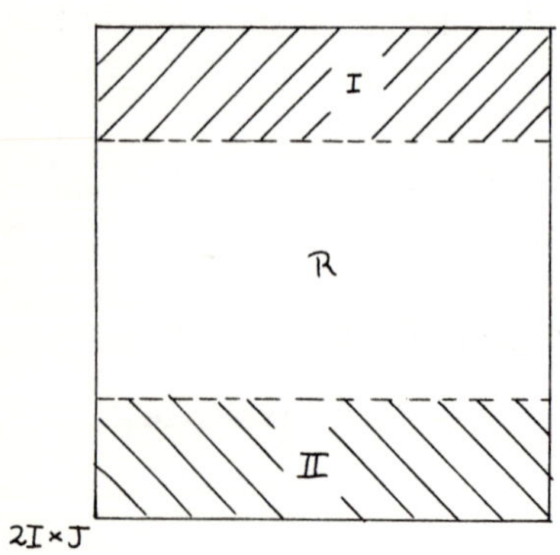

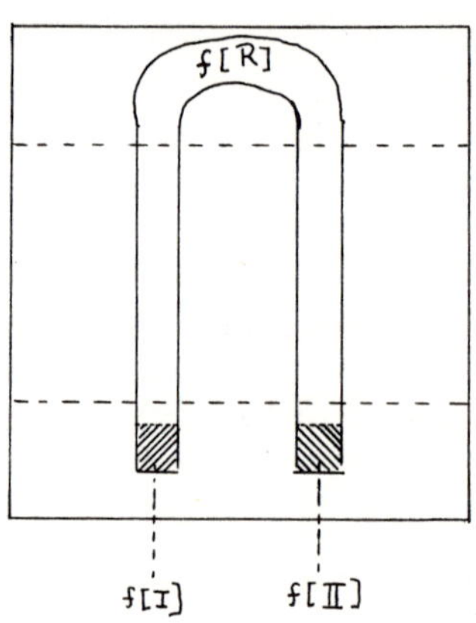

Such a description does not, of course, completely define f, but diffeomorphisms satisfying this description exist and any one of them will serve our purpose.

It then becomes clear that

(1) "II" contains a unique fixed sink
(2) "I" maps into "II"
(3) anything in R that eventually falls into "I" or "II" under either forward or backward iterates, never returns to R

so that, in treating the non-wandering set of f in R, we need only be concerned with points that stay inside R under all iterates of f. To keep track of where such points are traveling under f, we define subsets B_j^ℓ ($\ell \geq 0$, $j \geq -1$) of R inductively, via:

(1) $B_{-1}^{0} = R$

(2) $B_j^0 = f(B_{j-1}^0) \cap R$

(3) $B_{-1}^{\ell} = f^{-1}(B_0^0 \cap B_{-1}^{\ell-1})$

$\quad = f^{-1}(B_{-1}^{\ell-1}) \cap R$

(so that $B_{-1}^1 = R_1 \cup R_2$)

(4) $B_j^\ell = B_{-1}^\ell \cap B_j^0$.

We sketch the B_j^ℓ for low ℓ and j in figure 1. (It would probably be instructive for the reader to draw this sketch himself, directly from the definitions.)

Finally, for reasons which will soon become clear, we define B_{-1}^{∞}, B_{∞}^0, and B_{∞}^{∞} by taking intersections over all finite indices wherever the indicated index is infinite. We need to know two facts concerning the behavior of this family of sets.

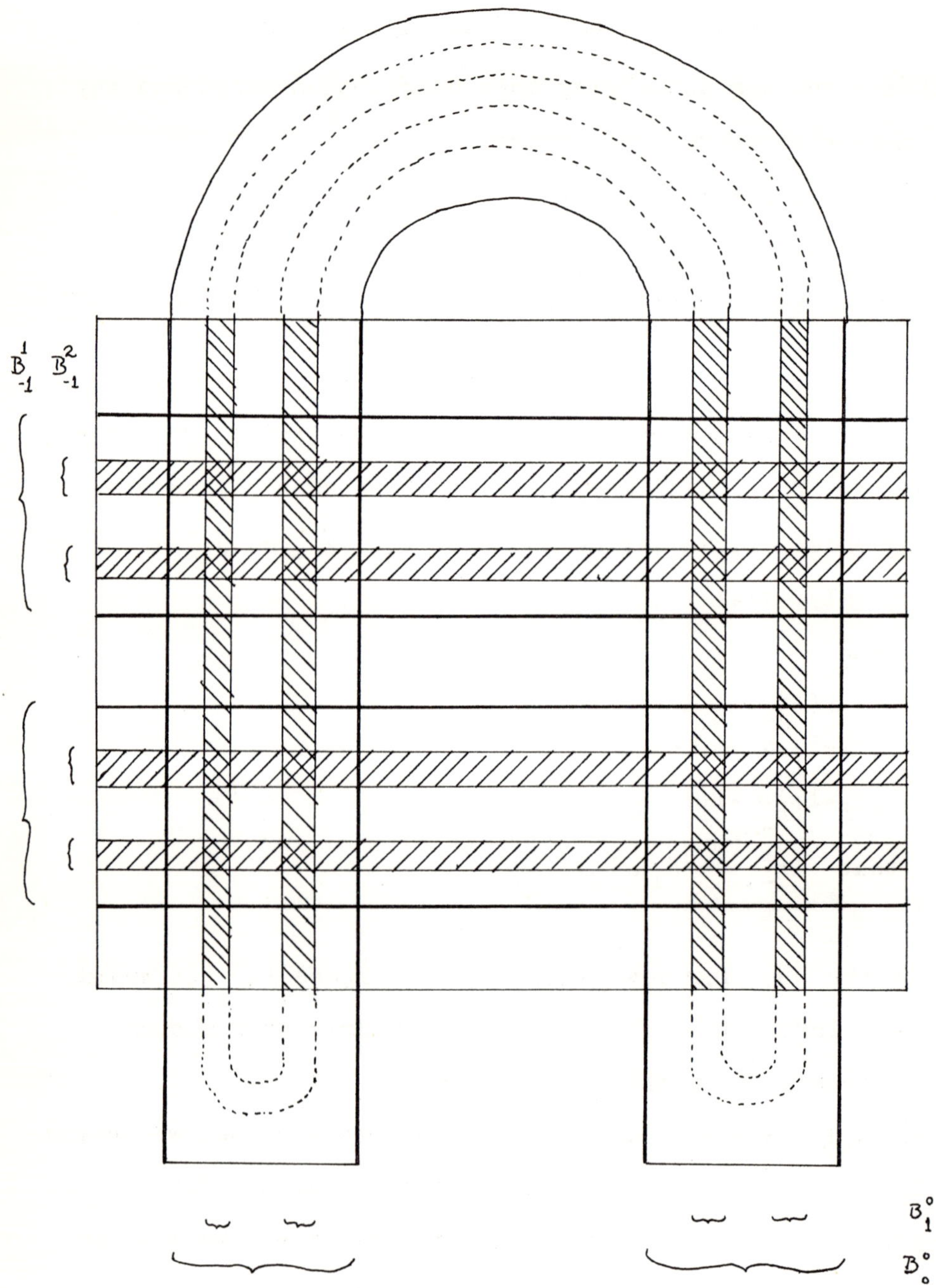

Figure 1. The low-numbered B^ℓ_j .

Lemma 2: For $\ell_0 \leq \ell \leq \infty$ and $j_0 \leq j \leq \infty$, $B_j^{\ell} \subset B_{j_0}^{\ell_0}$

(Remark that in view of this lemma in the finite case, we could view B_{-1}^{∞} as the inverse limit of the B_{-1}^{ℓ}, etc.)

Proof

By the definitions, it suffices to show the inequality with one of the indices fixed at its lowest possible value.

In the case of $\ell = \ell_0 = 0$, we want to show $B_j^0 \subset B_{j-1}^0$; this is clearly true (by inspection) for $j = 0$, and the induction step follows directly from (2) in the definition of the B_j^{ℓ}.

In the case $j = j_0 = -1$, we want to show $B_{-1}^{\ell} \subset B_{-1}^{\ell-1}$; the proof is similar. //

The infinite case then follows by taking intersections, or, equivalently, from an inverse limit argument.

Lemma 3: For $0 \leq \ell \leq \infty$ and $-1 \leq j \leq \infty$, $f|B_j^{\ell}$ is a homeomorphism onto $B_{j+1}^{\ell-1}$

Proof

The map f is a priori a homeomorphism; we need to show it is onto $B_{j+1}^{\ell-1}$, and the finite case (ℓ, $j < \infty$) will suffice.

To do this we stare at the definitions, and use the fact that f is a homeomorphism to factor through intersections:

$$\begin{aligned} f[B_j^\ell] &= f[B_j^o \cap B_{-1}^\ell] \\ &= f(B_j^o) \cap f[f^{-1}(B_{-1}^{\ell-1}) \cap R] \\ &= f(B_j^o) \cap B_{-1}^{\ell-1} \cap f(R) \\ &= f(B_j^o \cap R) \cap B_{-1}^{\ell-1} \\ &= f(B_j^o) \cap R \cap B_{-1}^{\ell-1} \qquad (\text{since } B_n^m \subset R \text{ for } (m,n) = (0,j) \\ &\qquad\qquad\qquad\qquad\qquad\qquad\qquad \text{or } (\ell-1,-1)) \\ &= B_{j+1}^o \cap B_{-1}^{\ell-1}. \qquad // \end{aligned}$$

We note that the B_j^ℓ are by no means connected sets: B_{-1}^1 and B_o^o each have two components (see the figure), B_{-1}^2 and B_1^o each have four, and so on. It is fairly clear, in fact, that each B_j^o consists of 2^{j+1} vertical strips (f "doubles up" each vertical strip) and B_{-1}^ℓ consists of 2^ℓ horizontal strips (for similar reasons). Hence B_j^ℓ, the collection of intersections in this "plaid" arrangement, consists of $2^{\ell+j+1}$ little rectangles scattered around R. If we are to have any real picture where points in R are going under f, we need to keep track of where these individual pieces are flying around.

We therefore define the set C_j^ℓ to be the collection of components of B_j^ℓ. For $\ell_o \leq \ell$, $j_o \leq j$, the inclusion map induces a map

$$i: C_j^\ell \longrightarrow C_{j_o}^{\ell_o}$$

That is, an element β of C_j^ℓ is a subset of R; since $B_j^\ell \subset B_{j_o}^{\ell_o}$, β is contained in some particular component of

$B_{j_o}^{\ell_o}$, which we can label $i(\beta)$.

On the other hand, the action of f on B_j^ℓ induces a map of components (which, by Lemma 3, is bijective)

$t: C_j^\ell \longrightarrow C_{j+1}^{\ell-1}$

This will be the machinery which enables is to analyze the orbit structure of f on R. We need one more fact:

<u>Lemma 4</u>: The diagram

$$\begin{array}{ccc} C_j^\ell & \xrightarrow{\quad t \quad} & C_{j+1}^{\ell-1} \\ i \big\downarrow & & \big\downarrow i \\ C_{j_o}^{\ell_o} & \xrightarrow[\quad t \quad]{} & C_{j_o+1}^{\ell_o-1} \end{array}$$

commutes, for $\ell_o \leq \ell$, $j_o \leq j$.

<u>Proof</u>

The lemma says, in more down-to-earth terms, that if f takes the component β of B_j^ℓ onto the component β' of $B_{j+1}^{\ell-1}$ and if β is contained in the component γ of $B_{j_o}^{\ell_o}$ then $\beta' = f[\beta]$ is contained in the component $f[\gamma]$ of $B_{j_o+1}^{\ell_o-1}$. But this is obvious. //

We close the section with the remark that all of these apparently finite arguments extend to the infinite case by taking appropriate intersections (= inverse limits).

4.3 FORMAL ANALYSIS OF $\Omega(f)$

We will now be concerned with analyzing the non-wandering set of f.

From the definition of f on $2I \times J$, we know that in this region, $\Omega(f)$ has a single isolated fixed sink (in "II") and that any other non-wandering points stay inside R under all iterates. We defined the B_j^ℓ in such a way that this set of points is just B_∞^∞.

Our analysis will take these steps: first, we will construct a map

$$h: B_\infty^\infty \longrightarrow A_\infty^\infty$$

which semi-conjugates $f|B_\infty^\infty$ with the shift automorphism. Since $\Omega(f) \subset B_\infty^\infty$ in R, restriction to $\Omega(f)$ will define a semi-conjugacy of $\Omega(f)$ to the shift. This semi-conjugacy will actually be constructed by defining a map on components,

$$\phi : C_\infty^\infty \longrightarrow A_\infty^\infty .$$

In turn, ϕ will be a bijection on C_∞^∞; we will show later that $\dim B_\infty^\infty = 0$, so that the "inclusion" map taking elements of B_∞^∞ into the components of B_∞^∞ (elements of C_∞^∞) to which they belong is an isomorphism $B_\infty^\infty \approx C_\infty^\infty$, and hence h is in fact a conjugacy. Then finally, the fact that $\Omega(\tau) = A_\infty^\infty$ and $\Omega(f) \cap R \subset B_\infty^\infty$ imply that $\Omega(f) \cap R = B_\infty^\infty$, and we will be done.

In this section we will be primarily concerned with constructing ϕ. To allow us later to apply this same analysis to perturbations of f, we take a somewhat axiomatic approach, deriving our analysis from certain properties which we will prove for f and its perturbation in 4.5. The only assumption we will need to make in this section is

<u>Property 1</u>: (1) R is connected

(2) C_0^0 is finite (say with m elements)

(3) If $\beta_1 \in C_{-1}^{\ell}$ and $\beta_2 \in C_j^0$ (ℓ, j finite) then $\beta_1 \cap \beta_2 \in C_j^{\ell}$ (ie, $\beta_1 \cap \beta_2$ is nonempty and connected).

We state parts (1) and (2) (which are obvious in our case) in their general form to point out that this analysis applies verbatim to somewhat different situations, as well. We will discuss the case when $m \neq 2$ later in this chapter. Part (3) will be established in 4.5.

We are concerned with finding a map ϕ making the diagram

$$\begin{array}{ccc} C_\infty^\infty & \xrightarrow{\ t\ } & C_\infty^\infty \\ \phi\downarrow & & \downarrow\phi \\ A_\infty^\infty & \xrightarrow[\ \tau\ ]{} & A_\infty^\infty \end{array}$$

commute, where t is the action of f on C_∞^∞ and τ is the shift automorphism.

Recall that one description of the shift automorphism was as an inverse limit of the system of maps

$$\begin{array}{ccc} A_j^{\ell} & \xrightarrow{\ \tau_j^{\ell}\ } & A_{j+1}^{\ell-1} \\ p\downarrow & & \downarrow p \\ A_{j_0}^{\ell_0} & \xrightarrow[\ \tau_{j_0}^{\ell_0}\ ]{} & A_{j_0-1}^{\ell_0-1} \end{array} \qquad (1)$$

Similarly, t is the inverse limit of the system of maps

$$\begin{array}{ccc} C^{\ell}_{j} & \xrightarrow{\;t\;} & C^{\ell-1}_{j+1} \\ {\scriptstyle i}\downarrow & & \downarrow{\scriptstyle i} \\ C^{\ell_0}_{j_0} & \xrightarrow[\;t\;]{} & C^{\ell_0-1}_{j_0+1} \end{array} \qquad (2)$$

Thus, to construct a bijection ϕ it will suffice to construct a bijection of inverse limit systems — that is, a collection of bijective maps

$$\phi^{\ell}_{j} : C^{\ell}_{j} \longrightarrow A^{\ell}_{j}$$

from the vertices of (1) to the vertices of (2) making the appropriate cube

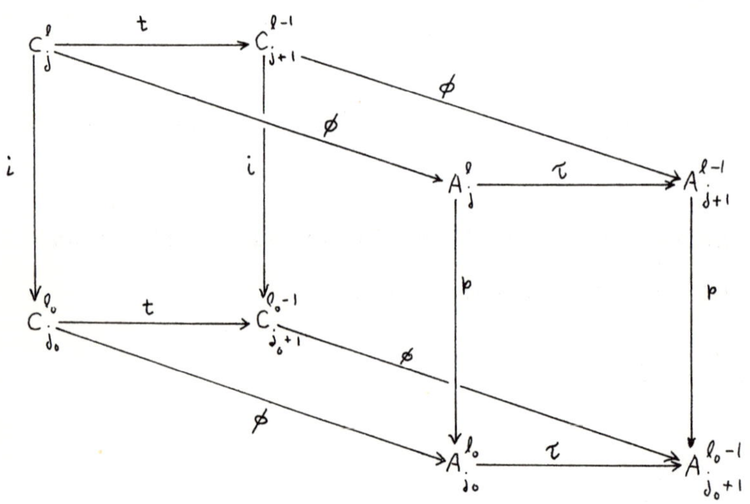

commute.

This is essentially a labelling problem, and takes on the flavor of an argument from homological algebra.

We have ϕ "naturally" defined on C^o_{-1} (to A^o_{-1}, which is non-existent, but could be defined arbitrarily as a one-element set) and on C^o_o (to $A^o_o = \mathcal{S}_m$). To extend by induction, we need a

<u>Key Lemma</u>: For $\beta_1, \beta_2 \in C^o_j$, the following statements are equivalent

(1) $f\beta_1 \cap \beta_2$ is nonempty

(2) $f\beta_1 \cap \beta_2 \in C^o_{j+1}$

(3) β_1 and $f^{-1}\beta_2$ are both contained in the same component of B^o_{j-1}.

<u>Proof</u>

First, we note, intuitively, that an element of some C^o_j is a vertical strip extending from the top to the bottom of R, so that its image will be a "thin" horseshoe, and the intersection of its image with R will be two strips, each extending from top to bottom of R. Thus, $f\beta_1 \cap R$ will consist of exactly two components of B^o_{j+1}; furthermore, one of these will be contained in $f[R_1]$ the other in $f[R_2]$—so that a single component of B^o_j (like β_2) can intersect at most one of these components, and in that case must actually contain it (since $B^o_{j+1} \subset B^o_j$). This shows the equivalence of (1) with (2).

To show (1) equivalent to (3), we note that, since f is bijective, $f\beta_1 \cap \beta_2 \neq \phi$ iff $\beta_1 \cap f^{-1}\beta_2 \neq \phi$. We will work with the latter condition. Again, intuitively, we know that $f^{-1}\beta_2$ is connected, and is in fact an element of C^1_{j-1}, so can be given as the intersection of R_1 or of R_2 (ie, an element of

C^1_{-1}) with an element of C^o_{j-1}. If we intersect $f^{-1}\beta_2$ with β_1, we see that $\beta_1 \cap f^{-1}\beta_2 \in C^1_{j-1}$. In particular, however, β_1 and $f^{-1}\beta_2$ are both connected subsets of B^1_j, so that each is contained in a unique element of C^1_j. Say $\beta_1 \subset \gamma_1$, $f^{-1}\beta_2 \subset \gamma_2$. Hence if they intersect, $\gamma_1 = \gamma_2$. Conversely, we see that $f^{-1}\beta_2 = \gamma_2 \cap R_i$ ($i = 1$ <u>or</u> 2) and $\beta_1 \cap R_i \neq \emptyset$ ($i = 1$ <u>and</u> $i = 2$) so that if $\gamma_1 = \gamma_2$, $\beta_1 \cap f^{-1}\beta_2 \neq \emptyset$. In the general case these intuitive arguments are direct consequences of part (3) of property 1. //

Now, to construct our semi-conjugacy, recall that we want to commute a cube with faces

$$\begin{array}{ccc} A^{\ell}_j & \xrightarrow{\tau} & A^{\ell-1}_{j+1} \\ \downarrow p & & \downarrow p \\ A^{\ell_o}_{j_o} & \xrightarrow[\tau]{} & A^{\ell_o-1}_{j_o+1} \end{array} \qquad\qquad \begin{array}{ccc} C^{\ell}_j & \xrightarrow{t} & C^{\ell-1}_{j+1} \\ \downarrow i & & \downarrow i \\ C^{\ell_o}_{j_o} & \xrightarrow[t]{} & C^{\ell_o-1}_{j_o-1} \end{array}$$

If we can construct ϕ so that the above cubes with $\ell_o = \ell$ or $\ell - 1$ and $j_o = j$ or $j - 1$ commute, we are done, since any other cube can be constructed from a finite number of these.

We first proceed to extend our definition from C^o_{-1}, C^o_o to C^o_j $(j > 0)$. We want ϕ to be functorial on the diagram

$$\begin{array}{ccccc} C^o_j & \xleftarrow{i} & C^1_j & \xrightarrow{t} & C^o_{j+1} \\ \downarrow i & & & & \downarrow i \\ C^o_{j-1} & \xleftarrow{i} & C^1_{j-1} & \xrightarrow[t]{} & C^o_j \end{array}$$

In particular, if we let

$$s_j : C_j^0 \longrightarrow C_{j-1}^0$$

be $i \circ t^{-1}$, this diagram reduces to

$$\begin{array}{ccc} C_j^0 & \xleftarrow{s_{j+1}} & C_{j+1}^0 \\ i \downarrow & & \downarrow i \\ C_{j-1}^0 & \xleftarrow[s_j]{} & C_j^0 \end{array}$$

Our induction step assumes ϕ_m^0 defoned for $m \leq j$; in particular, ϕ is defined on the three left-lower vertices of the last diagram.

On the other hand, by functoriality, we must have, for each $\beta \in C_{j+1}^0$,

$$\begin{array}{ccccc} (\phi(s_{j+1}\beta)) & \xleftarrow{p} & A_j^1 & \xrightarrow{t} & (\phi(\beta)) \\ p \downarrow & & \downarrow p & & \downarrow p \\ (\phi(s_j i\beta)) & \xleftarrow[p]{} & A_{j-1}^1 & \xrightarrow[t]{} & (\phi(i\beta)) \end{array}$$

Letting $\sigma_j = p \circ \tau^{-1}$, this says, for $\beta \in C_{j+1}^0$, that

$$p\ \sigma_{j+1}\ \phi(\beta) = \sigma_j\ p\ \phi(\beta)\ ,$$

and we must have

$$p\ \phi(\beta) = \phi(i\beta)$$

$$\sigma_{j+1}\phi(\beta) = \phi(s_{j+1}\beta)\ .$$

Hence, we must have

$$\phi(\beta) \subset \sigma_{j+1}^{-1}\ (\phi(s_{j+1}\beta)) \cap p^{-1}(\phi(i\beta))\ .$$

This last intersection contains at most a single element; it is nonempty provided $\sigma_j(\phi(i\beta)) \cap p(\phi(s_{j+1}\beta)) \neq \emptyset$. This last statement is equivalent to

$s_j(i\beta) \cap i(s_{j+1}\beta) \neq \emptyset$.

But the key lemma can be re-interpreted as saying

$$\beta = i^{-1}(i\beta) \cap s_{j+1}^{-1} \ (s_{j+1}\beta)$$

so that the assignment

$$\beta \longmapsto \sigma_{j+1}^{-1} \ (\phi(s_{j+1}\beta)) \cap p^{-1}(\phi(i\beta))$$

is well-defined and bijective.

Thus, we can lift the definition of ϕ to C_j^o , all j. Similar considerations show that by functoriality the diagram

$$\begin{array}{ccccc} C_{-1}^{\ell+1} & \xrightarrow{t} & C_o^{\ell} & \xrightarrow{i} & C_{-1}^{\ell} \\ \big\downarrow i & & & & \big\downarrow i \\ C_{-1}^{\ell} & \xrightarrow[t]{} & C_o^{\ell-1} & \xrightarrow[i]{} & C_{-1}^{\ell-1} \end{array}$$

determines $\phi(\beta) = p^{-1}(i\beta) \cap \tau^{-1} \ p^{-1} \ (i \ t \ \beta)$.

Finally, we can combine the above definitions to define ϕ on all C_j^{ℓ} by commuting the diagram

$$\begin{array}{ccc} C_{-1}^{\ell} \times C_j^o & \xrightarrow{\phi \times \phi} & A_{-1}^{\ell} \times A_j^o \\ \cap \big\downarrow & & \big\downarrow \quad (\text{"concatenation"}) \\ C_j^{\ell} & \xrightarrow[\phi]{} & A_j^{\ell} \end{array} .$$

4.4 LINEAR THEORY

We gather together in this section some facts from linear theory which will be useful in 4.5 for verifying property 1 and total-disconnectedness of B_{∞}^{∞} , for f and its perturbations.

Let $E = E_1 \oplus E_2$ be Banach spaces with norm given by

$$v^{2} = \| v_1 \|^2 + \| v_2 \|^2 .$$

<u>Definition</u>: A <u>horizontal</u> <u>sector</u> of size η ($\eta \in \mathbb{R}^+$) is a set of the form

$$s_\eta = \{(x, y) \mid \|y\| < \eta \|x\|\}$$

A <u>vertical</u> <u>sector</u> of size is a set

$$\bar{s}_\eta = \{(x, y) \mid \|x\| < \eta \|y\|\}$$

We are interested in the preservation of sectors by linear maps and their perturbations. A linear map $L: E \longrightarrow E$ can be written as a matrix

$$L = \begin{pmatrix} a & b \\ c & d \end{pmatrix}$$

of linear maps $E_i \longrightarrow E_j$. Assume $a: E_1 \longrightarrow E_1$ and $d: E_2 \longrightarrow E_2$ are invertible.

The basic fact we want is

<u>Lemma 5</u>: $L[s_\eta] \subset s_\eta$ provided

$$\frac{\|c\| + \|d\|\eta}{\|a\| - \|b\|\eta} < \eta$$

Similarly, $L[\bar{s}_\eta] \subset \bar{s}_\eta$ provided

$$\frac{\|b\| + \|a\|\eta}{\|d\| - \|c\|\eta} < \eta .$$

This tells in effect how badly L can fail to preserve the splitting if it is near a map that preserves the splitting.

<u>Proof</u>

We show the first case; the second clearly follows by reversing appropriate letters.

Note that $L(x,y) = (ax + by, cx + dy)$, so that we want to

have

$$\|cx + dy\| < \eta \|ax + by\| .$$

But (if $(x,y) \in s$)

$$\|cx + dy\| \leq \|c\|\cdot\|x\| + \|d\|\cdot\|y\|$$
$$\leq (\|c\| + \|d\|\eta)\|y\|$$

and

$$ax + by \geq (\|a\| - \|b\|\eta)\|x\|$$

and the estimate follows. //

We are primarily concerned with knowing when a perturbation of a hyperbolic linear map of a given skewness preserves a sector.

Corollary 1: For every μ, $0 < \mu < 1$, and every $\eta > 0$, there is $\varepsilon > 0$ such that any linear map T whose entries (in the matrix notation) satisfy the estimates

$$\begin{pmatrix} \geq \frac{1}{\mu+\varepsilon} & \leq \varepsilon \\ \leq \varepsilon & \leq \mu + \varepsilon \end{pmatrix}$$

(where each "$\leq$" means in norm, and the "$\geq$" means reciprocal of inverse norm,) will preserve s_η .

Note: The corresponding estimates for preserving $\bar{s}_\eta$ are

$$\begin{pmatrix} \leq \mu + \varepsilon & \leq \varepsilon \\ \leq \varepsilon & \geq \frac{1}{\mu+\varepsilon} \end{pmatrix} .$$

Proof

Plugging into the estimate of Lemma 5, we want $\varepsilon > 0$ so that for $\varepsilon' < \varepsilon$,

$$\frac{\varepsilon' + \eta(\mu+\varepsilon')}{\frac{1}{\mu+\varepsilon'} - \varepsilon'\eta} < \eta .$$

But the estimate clearly holds for $\varepsilon' = 0$ $\left(\frac{\eta\mu}{1/\mu} < \eta\right)$, so that by continuity it also holds for sufficiently small ε'. //

<u>Corollary 2</u>: If L_1, L_2 satisfy the above estimate and $c_0 > 0$, then

$L_1 \circ L_2$ and $c_0 \cdot L_1$

also preserve s_η. If L can be written as a product of matrices of the form $c_i \cdot L_i$ with $c_i > 0$ and L_i satisfying the estimates, then L preserves s_η.

The proof is obvious. We can now proceed to some "non-linear" analysis.

4.5 PERTURBATIONS OF f

In this section, we simultaneously complete the construction of the conjugacy

$$\Omega(f) \longrightarrow A_\infty^\infty$$

and prove Ω-stability of f, by verifying several properties for f and any nearby map g. We recall the two main properties of concern to us:

<u>Property 1</u>: (1) R connected

(2) C_0^0 has m elements

(3) If $\beta_1 \in C_{-1}^{\ell}$, $\beta_2 \in C_j^0$ $(\ell, j < \infty)$ then $\beta_1 \cap \beta_2 \in C_j^{\ell}$.

Property 2: $\dim B_\infty^\infty = 0$ (equivalently, B_∞^∞ is totally disconnected.

We point out that in reference to a perturbation g of f, the concepts "B_j^ℓ" and "C_j^ℓ" are still well-defined, so that it makes sense to ask if g possesses property 1 or property 2.

We start with some useful technical observations concerning f and g. Recall that by construction, f restricted to $R_1 \cup R_2$ is linear, and in fact hyperbolic linear. Thus, the tangent space over R_1 and R_2 has an invariant splitting, analogous to the decomposition over the whole manifold for Anosov diffeomorphisms.

To be more precise, R is a subset of the plane, so that the tangent space at any point can be canonically identified with $\mathbb{R}^2$. Under this identification we can identify subspaces E_1 and E_2 (they are, in fact, the vertical and horizontal lines) so that, if x is in $R_1 \cup R_2$, then fx is in R and the tangent map $Tf_{(x)}$ preserves the splitting, expands E_1 and contracts E_2. $Tf_{(x)}$ is in fact given by the matrix

$$\begin{pmatrix} \lambda_1^{(i)} & 0 \\ 0 & \lambda_2^{(i)} \end{pmatrix}$$

for $x \in R_i$ $(i = 1,2)$. Notice that this means also that $Tf|R_1 \cup R_2$ satisfies the inequalities of Lemma 5 and Corollary 1 (4.4) and therefore preserves any sector s_η, and $Tf^{-1}(x)$ preserves $\overline{s_\eta}$, for any $\eta > 0$. It is clear that for given

η , any sufficiently C^1-small perturbation g of f will have the same property. In particular, if $v \in E_1$, with $v \in T_y R$, and if $f^{j+1}y = x$ is an element of a component β of B^o_j (ie, if $y \in B^{j+1}_{-1}$) then

$Tf^{j+1}v \in S_\eta$

for any $\eta > 0$. Also, given η , a sufficiently C^1-small perturbation g will also satisfy

$Tg^{j+1}v \in S_\eta$

Now, to verify property 1 for perturbations g of f, we need a technical lemma to describe the sets B^ℓ_j obtained from g. To clarify notation, all of the sets from 4.2 which are derived from g instead of f will be denoted with a tilde: $\tilde{B}^\ell_j$, $\tilde{C}^\ell_j$, $\tilde{\beta}$, etc.

Lemma 6: If $\tilde{\beta} = \tilde{\beta}_1 \cap \tilde{\beta}_2$, with $\tilde{\beta}_1 \in \tilde{C}^o_j$ and $\tilde{\beta}_2 \in \tilde{C}^\ell_{-1}$ $(j, \ell < \infty)$ then (for g C^1-near f) there exists a diffeomorphism

$\psi_{\tilde{\beta}} : \tilde{\beta} \longrightarrow I \times J$

defined by

$\psi_{\tilde{\beta}}(x) = (\pi_1\, g^\ell(x) \times \pi_2\, g^{-(j+1)}(x))$.

Before giving a proof of the lemma, we point out how it gives us property 1:

Corollary 3: If g is C^1-near f, then g has property 1.

Proof of Corollary

The statement that g has property 1 means: we can take $\tilde{R}$ to be identical with the previous "R", then $\tilde{C}^o_o$ will have as many components as C^o_o (two), and a component $\tilde{\beta}_1$ of $\tilde{B}^o_j$ and $\tilde{\beta}_2$

of $\tilde{B}^{\ell}_{-1}$ intersect in a component of $\tilde{B}^{\ell}_{j}$.

It is obvious that any g C^0-near f will have two components for $\tilde{C}^0_0$. To show the last part, we note that the lemma says, in effect, that the components of $\tilde{B}^{\ell}_{j}$ inside $\tilde{\beta} \in \tilde{C}^{\ell_0}_{j_0}$ $(\ell_0 \leq \ell, j_0 \leq j)$ lie in $\tilde{\beta}$ in the same arrangement as the components of $\tilde{B}^{(\ell-\ell_0)}_{(j-j_0)}$ lie inside $\tilde{R}$. For f, then, property 1 clearly follows, since a component of B^0_0 intersects a component of B^{ℓ}_{-1} in a connected nonempty set; on the other hand, it is equally clear that for g C^1-near f, the "vertical" will stay well within a sector $\bar{s}_{\eta}$ and so again $\tilde{\beta}_1 \in \tilde{C}^0_0$ intersects $\tilde{\beta}_2 \in \tilde{C}^{\ell}_{-1}$ in an element of $\tilde{C}^{\ell}_0$, and we are done with the corollary. //

Proof of Lemma 6

We begin by picking η so that any element v of s_{η} makes an angle with E_1 (the vertical) of less than $\frac{\pi}{8}$.

(1) Claim: $\psi_{\tilde{\beta}}$ is a well-defined map:

We need to know, given $x \in \tilde{\beta}$, that $g^{\ell}(x), g^{-(j+1)}(x) \in R$. But by definition, $x \in \tilde{\beta}_1 \cap \tilde{\beta}_2$, $\tilde{\beta}_1 \subset \tilde{B}^0_j$, $\tilde{\beta}_2 \subset \tilde{B}^{\ell}_{-1}$, so $g^i(x) \in R$ $\quad (-(j+1) \leq i \leq \ell)$.

(2) Differentiability of $\psi_{\tilde{\beta}}$ follows from its definition.

(3) Claim: $\psi_{\tilde{\beta}}$ is non-singular:

Let L_1 denote the kernel of $T(\pi_1 g^{\ell})_x$ and L_2 the kernel of $T(\pi_2\, g^{-(j+1)})_x$. Then the kernel of $T\psi_{\tilde{\beta}}$ at x is $L_1 \cap L_2$. A vector v is in L_2 if and only if

$T(g^{-(j+1)})_x(v) \in E_1$, or in other words $v \in T(g^{j+1})_y[E_1]$, where $y = g^{-(j+1)}(x)$. By hypothesis, $T(g^{j+1})_y$ takes E_1 into s_η; hence $L_2 \subset s_\eta$.

On the other hand, similar arguments show $L_1 \subset \bar{s}_\eta$. Hence $\ker T\psi_{\tilde\beta} = L_1 \cap L_2 \subset s_\eta \cap \bar{s}_\eta$.

But the sectors with angle $\frac{\pi}{8}$ about the vertical and the horizontal meet only at the origin, so $\ker T\psi_{\tilde\beta}$ is trivial:

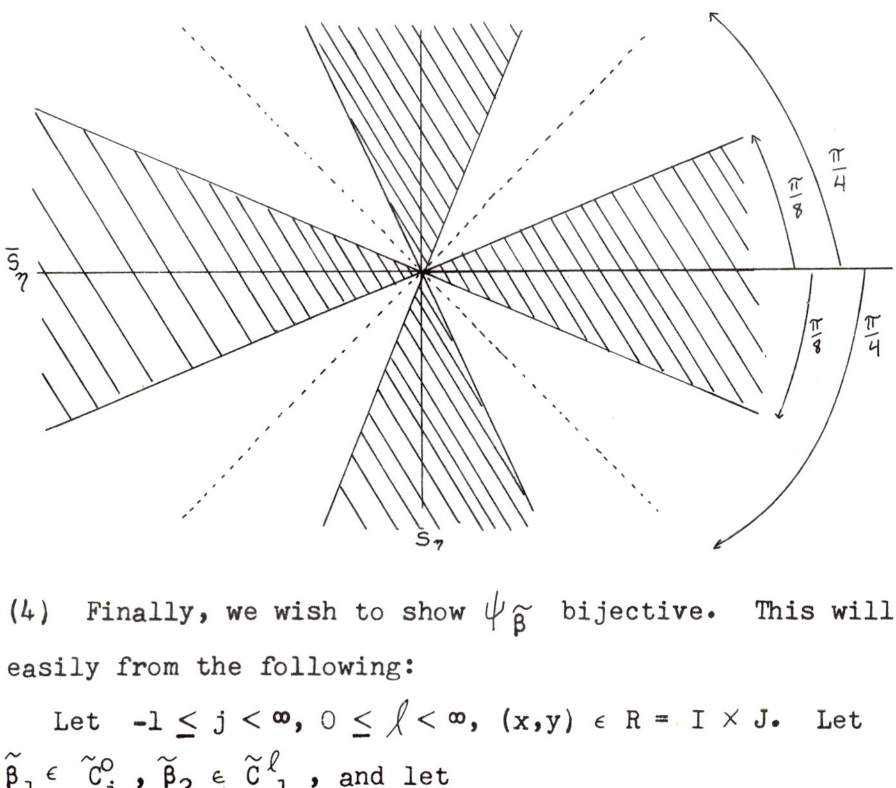

(4) Finally, we wish to show $\psi_{\tilde\beta}$ bijective. This will follow easily from the following:

Let $-1 \leq j < \infty$, $0 \leq \ell < \infty$, $(x,y) \in R = I \times J$. Let $\tilde\beta_1 \in \tilde{C}^0_j$, $\tilde\beta_2 \in \tilde{C}^\ell_{-1}$, and let

$E_x = \pi_1^{-1}(x) = x \times E_2$

$E_y = \pi_2^{-1}(y) = E_1 \times y$

(These are, respectively, the "horizontal" and vertical" lines through (x,y).)

Finally, let

$$W_1 = g^{-\ell}[E_x] \cap \tilde{\beta}_2$$
$$W_2 = g^{(j+1)}[E_y] \cap \tilde{\beta}_1 \quad .$$

Then we

<u>Claim</u>: $W_1 \cap W_2 = [\tilde{\beta}_2 \cap (\pi_1 g^{\ell})^{-1}(x)][\tilde{\beta}_1 \cap (\pi_2 g^{-(j+1)})^{-1}(y)]$ is a single point.

<u>Proof</u>

Inductively, we define

$$\tilde{B}^0_j(y) = g(\tilde{B}^0_{j-1}(y) \cap \tilde{B}^1_{-1}) \ ,$$

with $\tilde{B}^0_{-1}(y) = E_y$.

Then W_2 is a component of $\tilde{B}^0_j(y)$. By induction on j, $\pi_1 | W_2$ is a diffeomorphism with I (so that W_2 goes from top to bottom of R once) and $TW_2 \subset S_\eta$.

Similarly, we can define $\tilde{B}^{\ell}_{-1}(x)$; W_1 is a component, and so $\pi_2 | W_1$ is a diffeomorphism with J, and

$TW_1 \subset \bar{S}_\eta$.

But then the claim is obvious. //

Thus, by the arguments in 4.3, for any g C^1-near f, we have a semi-conjugacy

$$h_g : \tilde{B}^{\infty}_{\infty} \longrightarrow A^{\infty}_{\infty}$$

with the shift automorphism, defined as a conjugacy on the level of components:

$$\phi_g : \tilde{C}^{\infty}_{\infty} \longrightarrow A^{\infty}_{\infty} \ .$$

In particular, we have a bijection

$$\phi_g^{-1}\,\phi : \mathcal{C}_{\infty}^{\infty} \longrightarrow \tilde{\mathcal{C}}_{\infty}^{\infty} .$$

To show that this actually defines an Ω-conjugacy,

$$h_g^{-1}h: \ \Omega(f) \longrightarrow \Omega(g)$$

it will suffice to prove that f and g both have property 2.

We do this in a sequence of technical lemmas.

Lemma 7: If α is a compact, connected subset of $\bar{R}$ with the property that any two points $\delta, \gamma \in \alpha$ which differ by an element of a small sector

$$(\delta - \gamma) \in \bar{s}_{1/4}$$

must be within distance K of each other

$$|\delta - \gamma| < K$$

then any component α_1 of $g\alpha \cap R$ has the same property, but with $K/2$ instead of K.

Proof

Take p, q two points of α_1. Then we can write their difference as a column vector

$$p - q = \begin{pmatrix} \Delta x \\ \Delta y \end{pmatrix}$$

with $|\Delta x| \le \frac{1}{4} |\Delta y|$.

Let $\delta = g^{-1}p$, $\gamma = g^{-1}q$. Then by the Mean Value Theorem, there is an element ξ of α_1 such that

$$\delta - \gamma = g^{-1}p - g^{-1}q = Tg_{\xi}^{-1}(p-q)$$

which can be expressed in matrix form as

$$\begin{pmatrix} a & b \\ c & d \end{pmatrix}\begin{pmatrix} \Delta x \\ \Delta y \end{pmatrix} .$$

If we denote the difference of δ and γ by

$$\delta - \gamma = \begin{pmatrix} \hat{d}x \\ \hat{d}y \end{pmatrix}$$

then

$$|\delta - \gamma| \geq |\hat{d}y| \geq \| d \| \{1 - | \tfrac{c}{4d} |\} | \Delta y|$$
$$\geq 3 \; |\Delta y| .$$

Since $\bar{s}_{1/4}$ is invariant under Tg^{-1}, and $|\delta - \gamma| < K$, we have

$$|p-q| = \sqrt{|\Delta x|^2 + |\Delta y|^2}$$
$$= \sqrt{\frac{|\Delta y|^2}{4} + |\Delta y|^2}$$
$$= \frac{\sqrt{17}}{4} |\Delta y|$$
$$\leq \frac{\sqrt{17}}{4} |\delta - \gamma| < \frac{K}{2} . \quad //$$

Lemma 8: If $\tilde{\beta} \in \tilde{C}^0_\infty$, then π_1 is a diffeomorphism of $\tilde{\beta}$ with I. If $\tilde{\beta}$ $\tilde{C}^\infty_{-1}$, then π_2 is a diffeomorphism with J.

Proof

We already have the relevant maps onto, by arguments in lemma 6. To show them one-to-one, we note that, using lemma 7, if $\tilde{\beta} \in \tilde{C}^0_j$, the diameter of $\tilde{\beta} \cap E_x (x \in I)$ is $< 2^{-j}$. //

Lemma 9: If $\tilde{\beta}_1 \in \tilde{C}^0_\infty$ and $\tilde{\beta}_2 \in \tilde{C}^\infty_{-1}$, then $\tilde{\beta}_1 \cap \tilde{\beta}_2$ is a single point.

Proof

We can express $\tilde{\beta}_i$ as an intersection of components of finite $\tilde{B}^\ell_j$:

$$\tilde{\beta}_1 = \bigcap_{j=-1}^{\infty} \tilde{\alpha}_j \qquad \tilde{\alpha}_j \in \tilde{C}^o{}_j$$

$$\tilde{\beta}_2 = \bigcap_{\ell=0}^{\infty} \tilde{\alpha}^\ell \qquad \tilde{\alpha}^\ell \in \tilde{C}_j^\ell$$

so that

$$\tilde{\beta}_1 \cap \tilde{\beta}_2 = \bigcap_{i=1}^{\infty} (\tilde{\alpha}_i \cap \tilde{\alpha}^i)$$

But this intersection is nonempty by the finite intersection property, and has diameter approaching zero by arguments as in lemma 8. //

An immediate implication of lemma 9 is

Corollary 4: $\tilde{B}_{\infty}^{\infty}$ is totally disconnected (ie, any g near f has property 2.)

which in turn implies

Theorem 1: The non-wandering set of f inside R is conjugate to the shift automorphism on 2 symbols, and f is C^1 Ω-stable.

Proof

We retrace the major steps: By 4.3, we have bijections

$$\phi\colon C_{\infty}^{\infty} \longrightarrow A_{\infty}^{\infty}$$

$$\phi_g\colon \tilde{C}_{\infty}^{\infty} \longrightarrow A_{\infty}^{\infty}$$

which conjugate the action of $f(g)$ with τ. By Corollary 4, the "inclusion" maps

$$B_{\infty}^{\infty} \longrightarrow C_{\infty}^{\infty}$$

$$B_{\infty}^{\infty} \longrightarrow \tilde{C}_{\infty}^{\infty}$$

are actually homeomorphisms, so that composing, we get conjugacies

$h: B^\infty_\infty \longrightarrow A^\infty_\infty$

$h_g: \tilde{B}^\infty_\infty \longrightarrow A^\infty_\infty$.

Now, $\Omega(\tau) = A^\infty_\infty$, by lemma 1; therefore, $B^\infty_\infty = \Omega(f|R)$

$$\tilde{B}^\infty_\infty = \Omega(g|\tilde{R})$$

so that h and h_g are actually Ω-conjugacies

$h: \Omega(f|R) \longrightarrow A^\infty_\infty$

$h_g: \Omega(g|\tilde{R}) \longrightarrow A^\infty_\infty$.

In particular, h fits the first part of theorem 1. Also,

$h_g^{-1} \circ h : \Omega(f|R) \longrightarrow \Omega(g|\tilde{R})$

is an Ω-conjugacy of f and g. //

We close this section with a few observations on variations of the horseshoe. Recall that the analysis here would go through if C^0_0 were assumed to have $m > 2$ components instead of 2, except that we would use $\mathcal{S}_m$ instead of $\mathcal{S}_2$ to build up A^∞_∞. This situation can be achieved, as shown in the diagram, two different ways for $m = 3$:

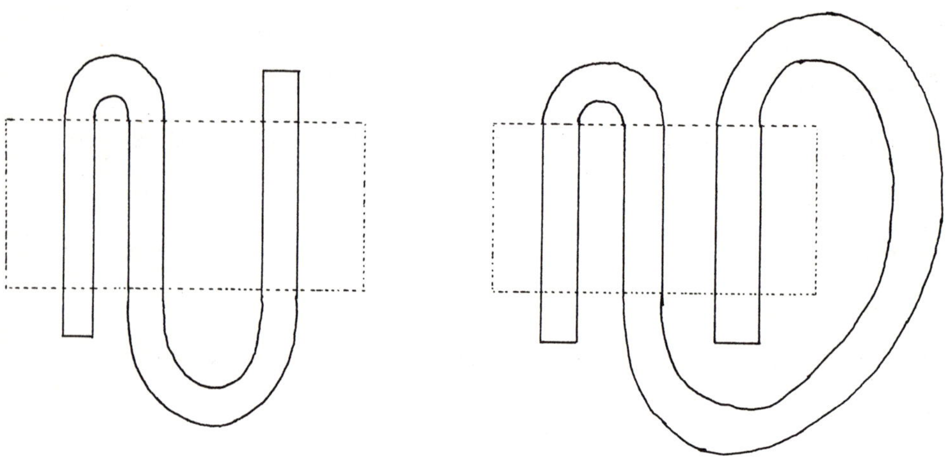

Thus, we can construct diffeomorphisms of compact surfaces which stably contain the shift on n symbols (any $n > 1$) as an "isolated" piece of the non-wandering set. The diagram on the right might arise, for example, if the image of R wraps around M in a homotopically non-trivial mod R way, for example if it is following an unstable manifold for an Anosov system. This analysis will be carried out in 4.6.

We can also extend the example to higher dimensions by replacing R with a hypercube and J_1, J_2 by disjoint balls on one face:

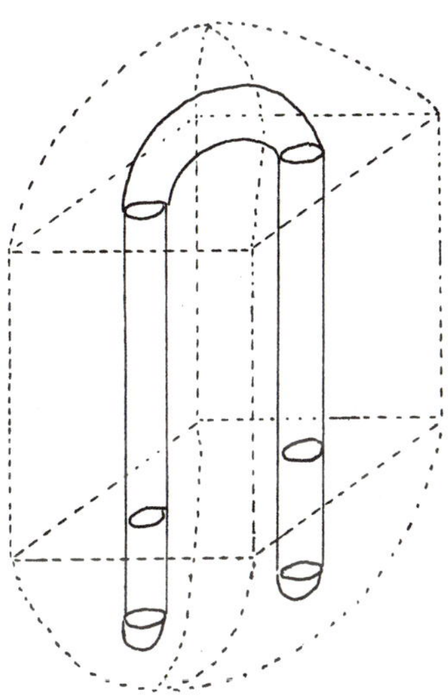

4.6 HOMOCLINIC POINTS

The homoclinic points are a phenomenon that actually occurs in nature, in the fullest sense of the word. They were originally pointed out in the context of the three-body problem by Poincaré ([1], vol III, pp. 377-383), who realized that the phenomenon involved tremendous difficulties: "The complexity of this figure will be striking, and I shall not even try to draw it." ([1], vol III, p, 382). For a recent analysis of such phenomena in the three-body problem, see Alekseev [1].

Definition: If p is a hyperbolic periodic point of f, then a point of intersection of $W^s(p)$ and $W^u(p)$ (other than p itself) is called a homoclinic point of p. If the intersection is transversal

$$W^s(p) \pitchfork_x W^u(p)$$

then x is called a transverse homoclinic point.

We note that a homoclinic point is the special case of the heteroclinic points of chapter 2 (p. 91) with the two periodic points identical. The definition of x homoclinic is equivalent to being doubly asymptotic, ie, for some m

$$\lim_{n \to +\infty} f^{mn} x = \lim_{n \to +\infty} f^{mn} x .$$

For our purpose, all homoclinic points will henceforth be assumed transverse. This is true in particular, whenever f is Kupka-Smale.

Poincare noticed that the existence of one homoclinic point implied the existence of an infinity of homoclinic points, with distinct orbits. We note, roughly, that since the stable and unstable manifolds of p are invariant, the orbit of a homoclinic point consists of homoclinic points; this forces

$W^u(p)$ to oscillate wildly as it approaches p, so that $W^u(p)$ piles up on itself, and similarly for $W^s(p)$, forcing a kind of "grid" of heteroclinic points.

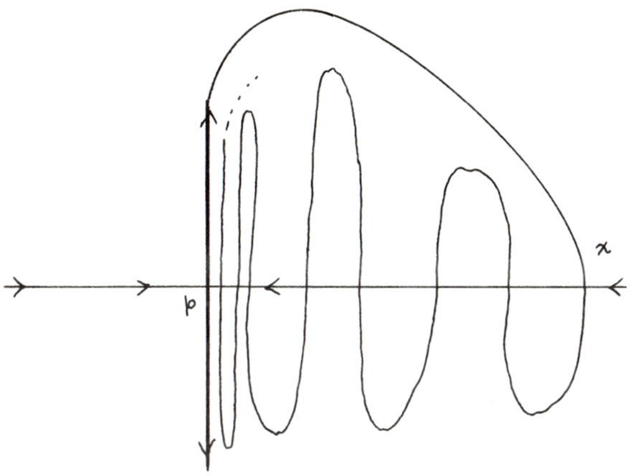

(1) $W^u(p)$ oscillates

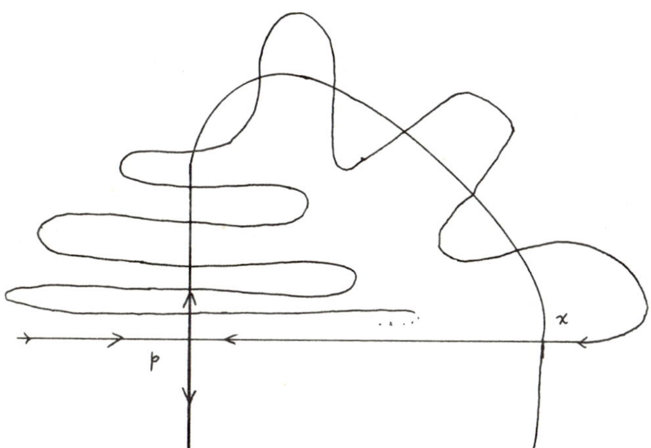

(2) $W^s(p)$ oscillates

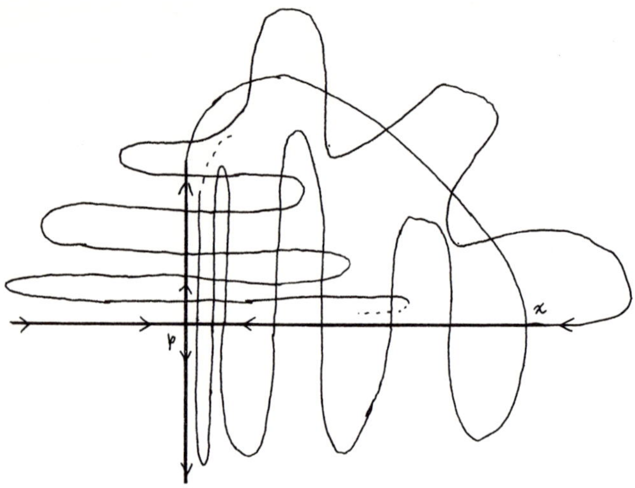

(3) So we have a grid of intersections.

Birkhoff studied this phenomenon further, and proved that every (transverse) homoclinic point is a limit of periodic points. In [4], Smale carries the analysis further, with the following theorem, which he says was probably known by Birkhoff.

Theorem 2 (Smale-Birkhoff): If f is Kupka-Smale, and x a homoclinic point of $p \in per(f)$, then there is a closed subset Λ of $\Omega(f)$, containing x, satisfying

(1) Λ is a cantor set

(2) $f\Lambda = \Lambda$

(3) $f^m|\Lambda \sim \tau | A_\infty^\infty$, for some m

Proof

As usual, we can assume p fixed. The essential point of the proof is to notice the similarity of a piece of the above figure to the horseshoe, and to then exploit our previous analysis.

By Hartman's theorem, f in a neighborhood of p is conjugate to a linear map L_o. We can assume the neighborhood can be written

$$V = B_{r_o} \times B_{r_1}$$

where the B_{r_i}'s are discs in E^u, E^s respectively. We can also assume that under this splitting, L_o can be written

$$L_o = \begin{pmatrix} A_1 & 0 \\ 0 & A_2 \end{pmatrix}$$

where $A_i : B_{r_i} \longrightarrow E_i$ is a linear map satisfying

$$\| A_1^{-1} \| < u < 1$$

$$\| A_2 \| < u < 1 .$$

Then, since $x \in W^s(p)$, some iterate of x is in V. If we carry out our analysis for this iterate, we can then use an iterate of f to conjugate the situation inside V with that around x. Hence, we assume $p, x \in V$.

An analogue of B_o^o is given by

Lemma 10: There exist (homeomorphs of) discs $K \subset W^s(p)$, $J \subset W^u(p)$, such that

(1) $x \in K \cap J$,

(2) $\partial K \cap J = \emptyset = K \cap \partial J$

(3) $K \subset$ interior $(f^{-1}[K])$

$J \subset$ interior $(f[J])$.

The proof is very clear from a straightforward application of stable and unstable manifold theory.

Now, since $W^s(p) \pitchfork W^u(p)$, $K \pitchfork J$, so that their intersection consists of a finite number of points, say $K \cap J = \{p_0, p_1, \ldots, p_r\}$.

By an application of the finite intersection property, assumption (3) above implies that $p \in K \cap J$; in effect, $p_1 \ldots p_r$ are the points of intersection which occur "before" x as we move out along W^s and W^u from p. We could (and shall) assume that $J \subset V$. Since the p_i move toward p under forward iterates, each is eventually in V. In particular, we can pick N large so that all iterates beyond the $N^{\underline{th}}$ for all the p_i lie in V: If $n \geq N$, then $(\forall\, i \leq r)$ $f^n(p_i) \in V$

Note that what is happening as we iterate p_i is that we move in along $J \subset W^s$, but out along W^u, and in particular out of K. Having picked N, let L_i denote the derivative of Df^N at p_i $(i=1\ldots r)$. Each L_i has the form

$$L_i = \begin{pmatrix} A & B \\ C & D \end{pmatrix}$$

with A invertible by the transversality of the homoclinic points. Pick an "error" $\varepsilon > 0$ so that if S is a matrix within ε of one of the L_i

$$\| S - L_i \| < \varepsilon$$

then S_1 is invertible, where S is expressed as

$$S = \begin{pmatrix} S_1 & S_2 \\ S_3 & S_4 \end{pmatrix} .$$

By continuity of the derivative, we can pick neighborhoods U_i of p_i so that for any $y \in U_i$, $f^N(y) \in V$ and

$\| Df^N(y) - L_i \| < \varepsilon$

We want to show that an iterate of f on V has properties 1 and 2 of 4.5. To do this, we want to invoke the analysis in 4.5. There are two steps. First,

Lemma 11: There exist integers $\lambda > 0$ and k_o such that for any $k > k_o$, $Df^{K+N+\lambda}(z)$ satisfies the estimates of corollary 2 (so that S_η is invariant, some small η) for each z with the property that the iterates

$f^{N+\lambda}(z)$, $f^{N+\lambda+1}(z), \ldots,$ $f^{N+\lambda+k}(z)$

are all in V and

$z \in f^{-\lambda}(U_i)$.

Proof

We note that for a z near $p_o = p$, $Df^{K+N+\lambda}(z)$ can be written

$Df^{K+N+\lambda}(z) = L_o^K \circ S \circ L_o^\lambda$, with S near L_o,

so that S_1 is invertible. But using the matrix notation

$$L_o^K \circ S \cdot L_o^\lambda = \begin{pmatrix} A_1^K S_1 A_1^\lambda & A_1^K S_2 A_2^{-\lambda} \\ A_2^K S_3 A_1^\lambda & A_2^K S_4 A_2^\lambda \end{pmatrix}$$

with S_1 invertible and $\| A^{-1} \|$, $\| A_1 \| < 1$. It is then clear that by picking λ and k_o large, we can blow up

$\| (A_1^K \circ S_1 \circ A_1^\lambda)^{-1} \|^{-1} = \| A_1^{-\lambda} \circ S_1^{-1} \circ A_1^{-K} \|^{-1}$

and cut down the other three entries. Now, picking λ, K_o valid for all the U_i, we let $U = \bigcup_{i=1}^{r} f^{-\lambda}U_i$ and let

$D_r = \{x_2 \in E_2 \mid \| x_2 \| \leq r \leq r_1\}$

be the r-disc in E_2. Our final statement, which is relatively easy, is

<u>Lemma 12</u>: There exist $r > 0$ and $k \geq k_o$ so that

$$f^{-\lambda}(K \cap J)$$

$$\subset f^{-(K+N+\lambda)} (f^{-\lambda}J \times D_r) (f^{-\lambda}J \times D_r)$$

$$\subset U .$$

Then it is clear that

$$f^{(K+N+\lambda)} \mid f^{-(K+M+\lambda)} (f^{-\lambda}J \times D_r)$$

satisfies properties 1 and 2, completing the theorem. //

We note finally that, besides the three-body problem, homoclinic points arise in any Anosov diffeomorphism, since it is clear that the stable and unstable manifold of a periodic point are dense. Thus, a full analysis of Anosov systems would also involve a good understanding of the horseshoe and shift. It can in fact be shown that the cantor set A_{∞}^{∞} and its shift can be mapped equivariantly (but clearly not homeomorphically) onto the torus and a hyperbolic automorphism.

CHAPTER 5

HYPERBOLIC SETS

While the theory of chapter 2 gives us a rather complete picture of the "generic" behavior of periodic points, the examples of the last two chapters make it clear that mere local analysis of periodic points does not give adequate insight into the global structure of diffeomorphisms. In this chapter, motivated by the Anosov and "horseshoe" examples, we try to formulate what we can of an analogous theory. While there are no analogues of the Hartman or Kupka-Smale theorems for hyperbolic sets in general, significant features of the old theory do, nevertheless, carry through.

The bulk of our effort will be the proof of the stable manifold theorem of Hirsch and Pugh [1] which includes as special cases the existence of the stable and unstable manifolds through a fixedpoint (p. 89) and of the stable and unstable foliation of an Anosov map (mentioned in Chapter 3). The proof (due to Hirsch and Pugh) also gives us the smoothness of all these manifolds, a fact we have not previously established. We will also prove the analogue of the stability theorem for hyperbolic fixedpoints (Theorem 3, Chapter 2). This is a step toward establishing the Ω-Stability theorem of Smale, which we will explore in chapter 6.

5.1 DEFINITION AND THE STABILITY THEOREM

The original idea of a "hyperbolic structure" was conceived by Perron, but the explicit formulation was first given by Smale ([3]). We have met the notion of hyperbolicity already, in several guises: hyperbolic linear maps, hyperbolic fixedpoints, hyperbolic toral automorphisms (and Anosov systems in general) and finally in the horseshoe. The idea was a bit less explicit in this last example, but recall that the map restricted to the inverse image of the rectangle $(B^{1}{}_{-1})$ was linear and hyperbolic, so that as long as the iterates of a point in R stayed in R, the iterate was linear and hyperbolic on a neighborhood. We can't of course hope to find linearity in general, but we can still try to formulate a concept of hyperbolicity for the derivative. The formulation is suggested by the definition of Anosov maps.

Let M be a manifold with a Riemannian metric, $f \in \mathrm{Diff}^{r}(M)$, and Λ a closed invariant set for f (ie, $f[\Lambda] = \Lambda$).

Definition: a hyperbolic structure for Λ is a splitting of the tangent bundle to M over Λ (the union of M_x for $x \in \Lambda$, denoted $T_\Lambda M$) into a continuous Whitney sum

$$T_\Lambda M = E^s \oplus E^u$$

satisfying the following conditions:

(1) Tf preserves E^s and E^u

(2) Tf contracts E^s, in the sense that for some uniform constants $c > 0$ and $0 < \lambda < 1$,

$$\| Tf^n v \| \leq c\, \lambda^n \| v \|$$

for every $v \in E^s$

(3) Tf expands E^u, in the sense that

$$\| Tf^n v \| \geq c\lambda^{-n} \| v \|$$

for every $v \in E^u$.

It should be pointed out that the continuity of the splitting does not imply that the dimension of E^s need be constant over all of Λ; different components of Λ might have different dimensions.

Clearly, as a special case, when Λ is a single point or a periodic orbit, this is just the description of a hyperbolic periodic point given by the lemma on p. 71; when $\Lambda = M$, this is the definition of an Anosov diffeomorphism.

It takes a bit more work to decide that the cantor set in the horseshoe example has a hyperbolic structure, but this is essentially what our main concern was in 4.4 and part of 4.5 in the last chapter.

Another example of a set with hyperbolic structure, this time definitely not consisting of non-wandering points, can be constructed by taking a transverse heteroclinic point x (for hyperbolic fixedpoints p and q) and taking as our splitting over the orbit of x that already given by $W^s(p) \pitchfork W^u(q)$. This splitting can be extended to the closure of $O(x)$.

Several quick observations follow from arguments we have seen before. Mather's estimates (given on p. 73) can be applied to any compact set with hyperbolic structure to show that under an equivalent metric, we can take $c = 1$. Such a metric is said to be <u>adapted</u> to the set. Conversely, when Λ is compact, the existence of a hyperbolic structure on Λ is independent of the choice of metric. We can thus refer to a compact, invariant set for f with hyperbolic structure as simply a <u>hyperbolic set</u> for f.

The relation of hyperbolic sets to hyperbolic fixedpoints is further pointed out by the following generalization of lemma 3 chapter 3 (p. 104):

<u>Lemma 1</u>: Consider the space $\mathcal{M}(\Lambda, M)$ of maps (not necessarily continuous) of Λ into M. Let F denote the operator taking $\varphi \in \mathcal{M}(\Lambda, M)$ into

$$f \circ \varphi \circ h^{-1} = F(\varphi)$$

where h is the automorphism of Λ given by the restriction of f to Λ.

Then Λ is a hyperbolic set for f if and only if the usual inclusion

$$i : \Lambda \hookrightarrow M$$

is a hyperbolic fixedpoint of F.

The proof is just a translation of lemma 3 (p. 104) to the case of a proper subset of M, and is left as an exercise for the reader. As before, there is also a continuous version, dealing with the subspace $C^0(\Lambda, M)$ of $\mathcal{M}(\Lambda, M)$.

A consequence of this fact is a weak analogue of theorem 3, chapter 2, and of the Anosov stability theorem (theorem 1, chapter 3). It also partly generalizes the stability of the horsehsoe (theorem 1, chapter 4).

<u>Theorem 1</u> (Stability of Hyperbolic Sets): If Λ is a (compact) hyperbolic set for $f \in \mathrm{Diff}^r(M)$ and g is a sufficiently C^1-small perturbation of f, then there exists a compact set, Λ_g, invariant for g, and a conjugating homeomorphism φ, uniformly near the inclusion map:

$$\begin{array}{ccc} \Lambda & \xrightarrow{\ f\ } & \Lambda \\ \varphi \downarrow & & \downarrow \varphi \\ \Lambda_g & \longrightarrow & \Lambda_g \end{array}$$

<u>Proof</u>

The proof is a direct re-interpretation of the Anosov stability theorem.

A conjugating map φ is a solution of the functional equation

$\varphi = g \circ \varphi \circ h^{-1}$

where h, as before, is $f|\Lambda$. This can be reformulated by asking that φ be a fixedpoint of the operator

$\mathcal{H}_g : C^0(\Lambda, M) \longrightarrow C^0(\Lambda, M)$

defined by

$\mathcal{H}_g(\varphi) = g \circ \varphi \circ h^{-1}$.

But $\mathcal{H}_f(\varphi) = f \circ \varphi \circ h^{-1}$ has a hyperbolic fixedpoint at the inclusion map, and $\mathcal{H}_g$ is a C^1-perturbation of $\mathcal{H}_f$, so that $\mathcal{H}_g$ has a fixedpoint φ C^0-near $i: \Lambda \hookrightarrow M$.

We then use the expansiveness of a hyperbolic set (another direct translation: lemma 5, chapter 3) and the claim at the top of p. 111 to show that φ is a homeomorphism of Λ with $\varphi[\Lambda]$, which we christen Λ_g. This completes the proof. //

We note that the stability demonstrated here for hyperbolic

sets is weaker than the local stability shown for points in chapter 2. There, the conjugacy was actually defined on a neighborhood of the point, whereas here, it is restricted to the set itself. Thus, while Λ may be "isolated" in the sense that the only invariant sets near Λ must include it, Λ_g need not have the corresponding property: hyperbolic sets can, and in some sense do, "explode", as we shall see in the next chapter. However, one could formulate proposition 1 as the statement that hyperbolic sets cannot collapse, or "implode."

5.2 STABLE MANIFOLD THEORY FOR A POINT

Having established some general, "global" properties of hyperbolic sets, we proceed to a more careful analysis of their "local" structure. Such an analysis was essentially given for fixedpoints by Hartman's theorem. While an analogous theorem was formulated and proved by Hirsch, Pugh, and Shub [1] for the case of an embedded invariant submanifold, the case of general hyperbolic sets (eg, the cantor set in the horseshoe) seems difficult to even formulate, lacking a candidate to replace the linear map $(Df)_p$.

An important feature of the analysis at a point is the existence of a stable (and unstable) manifold, and this part of the analysis can be generalized. The basic treatment, first given by Hirsch and Pugh [1], has two parts. First (in this section), we will give a direct proof (without Hartman's theorem) of the existence of a stable (and unstable) manifold for a hyperbolic fixedpoint of a locally defined diffeomorphism in a Banach space; this proof will simultaneously establish the smoothness of these manifolds. Then (in the next section), we will treat the case of a compact hyperbolic set Λ by lifting to a fixedpoint in $\mathfrak{M}(\Lambda, M($ (via lemma 1) and applying the fixedpoint case.

Recall a few definitions from chapter 2. A linear isomorphism T of Banach space E is <u>hyperbolic</u> iff its spectrum lies off the unit circle. If U is an open subset of E and $f: U \longrightarrow E$ a diffeomorphism with its image, then $p \in U$ is a <u>hyperbolic fixedpoint</u> of f if $fp = p$ and $(Df)_p$ is hyperbolic. In the linear case, recall that we had an invariant decomposition

$$E = E_1 \oplus E_2$$

into an expanding and contracting part. We call

$$\tau = \max(\|T_2\|, \|T_1^{-1}\|)$$

the <u>skewness</u> of T (where $T_i = T|E_i$) and

$$m(T_1) = \|T_1^{-1}\|^{-1}$$

the <u>minimum norm</u> of T_1.

We also note two "technical" facts whose proof is intuitively clear, and will be left to the reader.

<u>Proposition 2:</u> If T is a linear hyperbolic isomorphism with associated splitting $E = E_1 \oplus E_2$ then

(1) E_i can be characterized by:

$$E_1 = \{x \in E \mid \|T^{-n}x\| \longrightarrow 0 \text{ as } n \longrightarrow +\infty\}$$
$$= \{x \in E \mid \|T^{-n}x\| \text{ stays bounded for } n \geq 0\}$$
$$E_2 = \{x \in E \mid \|T^{n}x\| \longrightarrow 0 \text{ as } n \longrightarrow +\infty\}$$
$$\{x \in E \mid \|T^{n}x\| \text{ stays bounded for } n \geq 0\}$$

(2) If V is a bounded neighborhood of 0 in E_1 with $T[V] \supset V$, then for any uniform neighborhood W of $T[V] \sim V$ in E,

$$\left(\bigcup_{n \geq 0} T^{-n}W\right) \cup E_2$$

is a neighborhood of E_2.

The second statement can best be understood with a "picture"; it says that certain kinds of open sets about E_1 "pile up" on E_2 under backward iteration.

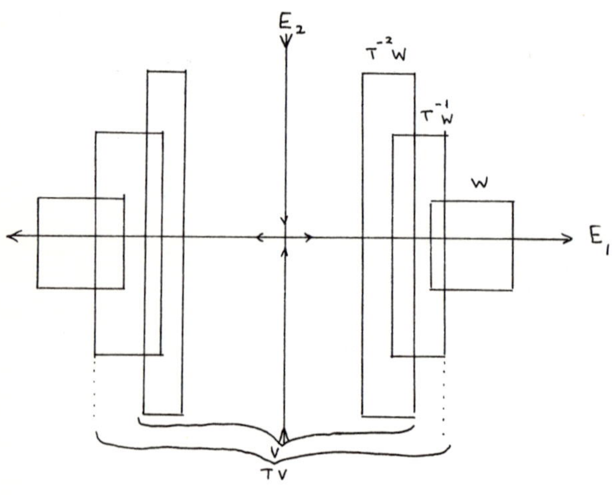

For the rest of this section, we fix a linear hyperbolic isomorphism T with skewness $\tau < 1$ (possibly after re-norming) and minimum norm $m(T_1) = \|T_1^{-1}\|^{-1}$. A technical lemma analogous to the analysis in 4.4 is

Lemma 2: If $f: E(r) \longrightarrow E$ is an ε-Lipschitz perturbation of T with $\varepsilon < \frac{1-\tau}{1+\tau}$, then for any pair $x, y \in E$ of vectors differing by an element of the sector s_1 (ie, if $\|x_1-y_1\| \geq \|x_2-y_2\|$) their images also differ by an element of the sector s_1; in fact,

$$\begin{aligned} \|f_1x - f_1y\| &\geq (\tau^{-1}-\varepsilon) \quad \|x_1-y_1\| \\ &\geq (\tau + \varepsilon) \quad \|x_1-y_1\| \\ &\geq \|f_2x - f_2y\| . \end{aligned}$$

where $f_i x$ denotes the projection of fx into E_i.

Proof

This is a fairly straightforward series of estimates. Clearly,

$$\begin{aligned} \|f_1x - f_1y\| &= \|T_1(x-y) + (f_1-T_1)(x) - (f_1-T_1)(y)\| \\ &\geq \|T_1(x_1-y_1)\| - \varepsilon \|x-y\| \end{aligned}$$

and since $\|x-y\| = \max_{i=1,2} \|x_i-y_i\| = \|x_1-y_1\|$, this last term is

$$\geq (\tau^{-1} - \varepsilon) \quad \|x_1-y_1\| .$$

Now, since $\varepsilon < \frac{1-\tau}{1+\tau}$, we have the pair of inequalities

$$\varepsilon < 1 - \tau$$

$$\varepsilon < \tau^{-1} - 1$$

in other words,

$$\tau + \varepsilon < 1 < \tau^{-1} - \varepsilon .$$

Finally, the proof that

$$\|f_2x - f_2y\| \leq (\tau + \varepsilon) \quad \|x_1 - y_1\|$$

is similar to the case above, with f_1. //

As usual, our proof of the stable manifold theorem will consist of finding a fixedpoint for an appropriate operator.

The fixedpoint lemma we use is a "parametrized" version of the contraction lemma (lemma 1, p.69).

Lemma 3 (Fiber Contraction Lemma): Let X be a topological space, $f:X \longrightarrow X$ a continuous map with attractive fixedpoint p. Let Y be a metric space, and let $f:X\times Y \longrightarrow X\times Y$ be a fiber-bundle map covering f - that is, F is of the form

$f(x,y) = (f(x), g_x(y))$.

Suppose q is a fixedpoint of g_p .
Then (p,q) is an attractive fixedpoint for F, provided

$$\limsup_{n \to \infty} [\mathscr{L}ip\,(g_{f_x^n})] < 1 \qquad (\forall\, x \in X) .$$

Recall that a fixedpoint of a map is attractive iff it is the unique common limit of all forward orbits.

Proof

We can simplify the problem a bit, since by replacing X with the subspace $\{p\} \cup \{f^n(x) \mid n \in \mathbb{Z}\}$ we can assume a bound on $\mathscr{L}ip(g_x)$:

$\mathscr{L}ip(g_x) \leq \lambda < 1 \qquad (\forall\, x)$

and it will suffice to show that

$\pi_2\, F^n(x,y) \longrightarrow q$

where $\pi_2:X\times Y \longrightarrow Y$ is the projection.

We first show the special case of the above assertion when $y = q$, ie,

$\pi_2\, F^n(x,y) \longrightarrow q.$

Let d_n denote the distance (in Y) between

$\pi_2\, F^n(x,q)$ and q .

Let $\delta_n = d(g_{f_x^n}(q), q)$. Then $d_n \leq d(\pi_2 F^n(x,q), g_{f_x^{n-1}}(q)) + \delta_n$

which in turn (since $\mathcal{L}ip(g_x) < \lambda$) is

$\leq \lambda\, d_{n-1} + \delta_n$.

Thus by induction,

$$d_n \leq \sum_{j=0}^{n} \lambda^{n-j}\, \delta_j$$

$$= \sum_{j=0}^{K-1} \lambda^{n-j}\, \delta_j + \sum_{j=K}^{n} \lambda^{n-j}\, \delta_j$$

$$\leq (\lambda^n + \ldots + \lambda^{n-K+1}) \max_j \delta_j$$

$$+ (1 + \ldots + \lambda^{n-K}) \max_{j \geq K} \delta_j$$

which, in the limit as $n \longrightarrow +\infty$, becomes

$$\leq \frac{\lambda^{n-K}}{1-\lambda} \max_j \delta_j + \frac{1}{1-\lambda} \max_{j \geq K} \delta_j$$

which goes to zero as K grows. Now,

$d(\pi_2 F^n(x,y), q)$

$\leq d(\pi_2 F^n(x,y), \pi_2 F^n(x,q)) + d_n$

$\leq \lambda^n\, d(y,q) + d_n \longrightarrow 0$.

and we have shown (p,q) attractive. //

One more piece of machinery must be constructed before we proceed to the central theorem. Recall from our analysis in chapter 2 that we expect $W^s(p)$ and $W^u(p)$ to be immersed planes through p deviating relatively little (locally) from the projection of E^s and E^u ; we will be defining them locally, so that their invariance will be of a special kind: W^s will map into itself, and W^u will be "overflowing invariant" - ie, $W^u \subset fW^u$. Finally, we will want to establish differentiability.

The basic trick for dealing with all these questions will be to try to view W^s and W^u (locally) as the graphs of smooth maps $\varphi : E^s \longrightarrow E^u$ (resp. $E^u \longrightarrow E^s$) with very slight slope. The invariance will then have to take the form: $f(\mathrm{gr}\,\varphi) \cap R = \mathrm{gr}\,\varphi$ where R will be some region in which we will be considering the graph. To use this idea in conjunction with functional analysis for φ, we define the notion of a <u>graph transform</u>.

We will be concerned with maps $g: E_i(r) \longrightarrow E_j$, and especially their graphs within $E(r)$. The graph transform by f of g, denoted $\Gamma_f(g)$, will be quite naturally defined in terms of its graph:

$$\mathrm{gr}(\Gamma_f(g)) = f[\mathrm{gr}(g)] \cap E(r) \; . \tag{1}$$

To express this by a functional formula, recall that the graph of $g: E_1(r) \longrightarrow E_2(r)$ can be regarded as the map $(\mathrm{id}, g) : E_1(r) \longrightarrow E(r)$.

If we are trying to transform the graph of g by the linear map T, then the formula is very simply:

$$\Gamma_T(g) = T_2 \circ g \circ T_1^{-1} \mid E_1(r) \; .$$

In the case of more general f (in our case, Lipschitz perturbations of T), the definition (1) becomes

$$\Gamma_f(g) \circ f_1 \circ (\mathrm{id}, g) = f_2 \circ (\mathrm{id}, g)$$

Letting the two composites with f be denoted

$$\psi_f(g) = f_1 \circ (\mathrm{id}, g)$$
$$\varphi_f(g) = f_2 \circ (\mathrm{id}, g)$$

we see that the graph transform is defined in case $\psi_f(g)$ is invertible by

$$\Gamma_f(g) = \varphi_f(g) \circ \psi_f^{-1}(g) \tag{2}$$

It is probably instructive for the reader to trace these formulas through a diagram involving a real graph. It is good manners to formalize definitions:

Definition: If $g: E_1(r) \longrightarrow E_2(r)$ is a function and $f: E(r) \longrightarrow E$ a diffeomorphism, then the graph transform of g by f, $\Gamma_f(g)$, is defined by

$$f[gr(g)] = gr[\Gamma_f(g)]$$

whenever this formula makes sense.

We can now state the central theorem of this section. It turns out easier to deal with unstable manifolds, and then to deduce the stable theory by inverting the map.

Theorem 2 (Unstable Manifold Theorem for a Point): Given $0 < \tau < 1$, $r > 0$, there exist a constant ε (independent of r) and a constant δ, $\varepsilon > \delta > 0$, such that if T is a linear hyperbolic isomorphism of E with skewness τ and $f: E(r) \longrightarrow E$ is a Lipschitz perturbation satisfying

(a) $\| f(0) \| \leq \delta$
(b) $\mathcal{L}ip(f - T) < \varepsilon$

then there exists a unique map

$$g_f : E_1(r) \longrightarrow E_2(r)$$

such that

$$gr(g_f) = W^u = \bigcap_{n \geq 0} f^n(E(r)) .$$

Moreover, g_f has the following properties:

(1) $\mathcal{L}ip(g_f) \leq 1$

(2) g_f is as differentiable as f

$(f \in \mathrm{Diff}^r(E) \Longrightarrow g_f \in C^r(E_1, E_2))$

(3) the assignment

$$f \longrightarrow g_f$$

is a continuous map

$$\mathcal{N}^K_\delta(T) \longrightarrow C^K(E_1(r), E_2(r))$$

(where $\mathcal{N}^K_\delta(T)$ denotes the space of C^K maps satisfying (a) and (b) with δ replacing ε) in other words, g_f depends continuously on f in the C^K topology

(4) $(f|W^u)^{-1}$ is a contraction of W^u.

<u>Proof</u>

We will find g_f as a fixedpoint of the graph transform - that is, $gr(g_f) \subset f[gr(g_f)]$.

To simplify matters, we will begin by assuming $f(0) = 0$, and deal with the more general case at the end of the proof. Begin by picking

$$\varepsilon < \frac{1-\tau}{1+\tau} = \frac{\tau^{-1} - 1}{\tau^{-1} + 1}$$

Thus, ε will satisfy the inequalities

$$\varepsilon < 1 - \tau$$

$$\varepsilon < \tau^{-1} \tag{3}$$

$$\frac{\tau + \varepsilon}{1 - \tau\varepsilon} < 1$$

To see the last inequality, translate the defining inequality to read

$\varepsilon(1 + \tau) < 1 - \tau$

and hence

$\tau + \varepsilon < 1 - \tau\varepsilon$.

Denote by $\mathcal{G}$ the space of maps of $E_1(r)$ into $E_2(r)$ with Lipschitz constant ≤ 1 and taking the origin into the origin. Clearly, $\mathcal{G}$ is a complete metric space under the uniform sup metric. Our first major step will be to show that Γ_f is defined on $\mathcal{G}$, maps into $\mathcal{G}$, and is in fact a contraction.

The fact that for any $g \in \mathcal{G}$, $\Gamma_f(g)$ is a well-defined map, also in $\mathcal{G}$, follows directly from lemma 2. The fact that Γ_f contracts $\mathcal{G}$, together with the fact that the fixedpoint of Γ_f is the whole unstable manifold, is a consequence of the following fact, shown me by M. Shub:

<u>Claim</u>: If $g \in \mathcal{G}$ and $(x,y) \in E_1(r) \times E_2(r)$, then the vertical distance of $f(x,y)$ from $f(gr(g))$ is contracted.

We consider the situation geometrically: the point $(x, g(x))$ directly below (x,y) on $gr(g)$ will be taken under f to a point $(f_1(x, gx), f_2(x, gx))$ which will lie on $gr(\Gamma_f g)$ but not necessarily below $f(x,y)$. We estimate first the effect of f on vertical and horizontal distances between these two "known" points. For notational convenience, let

$$f(x,y) = (x_1, y_1) \tag{0}$$

Then

$$\begin{aligned}|x_1 - f_1(x, gx)| &\le |(f_1 - T_1)(x,y) - (f_1 - T_1)(x, gx)| \\ &+ |T_1(x,y) - T_1(x, gx)| \\ &\le \mathcal{L}ip(f_1 - T_1)\ |(x,y) - (x, gx)| + 0 \\ &< \varepsilon\ |y - g(x)| \qquad (1)\end{aligned}$$

$$\begin{aligned}|y_1 - f_2(x,gx)| &\le |(f_2 - T_2)(x,y) - (f_2 - T_2)(x,gx)| \\ &+ |T_2(x,y) - T_2(x, gx)| \\ &\le \mathcal{L}ip(f_2 - T_2)\ |(x,y) - (x, gx)| \\ &+ \mathcal{L}ip\ T_2\ |y - gx| \\ &< \varepsilon|y - gx| + \tau|y - gx| \\ &= (\tau + \varepsilon)\ |y - g(x)| \qquad (2)\end{aligned}$$

Now, the point directly below $f(x,y) = (x_1, y_1)$ on $f(gr(g))$ is $(x_1, \Gamma_f g(x_1))$. The vertical distance of $f(x,y)$ to $f(gr(g))$ is then given by

$$\begin{aligned}|y_1 - \Gamma_f(g)(x_1)| &\le |y_1 - \Gamma_f g(f_1(x,gx))| \\ &+ |\Gamma_f g(f_1(x,gx)) - \Gamma_f g(x_1)| \\ &\le |y_1 - f_2(x, gx)| + [\mathcal{L}ip\ \Gamma_f g]|f_1(x,gx) - x_1| \\ \text{(by (1) and (2))}\quad &< (\tau + \varepsilon)|y - g(x)| + 1(\varepsilon|y - g(x)|) \\ &= (\tau + 2\varepsilon)|y - gx|\ .\end{aligned}$$

With this fact in hand, we easily see that any two graphs from g are contracted under f (the sup norm is just the sup of vertical distances); thus Γ_f has a unique, attractive fixedpoint, $g_f \in \mathcal{G}$. Clearly, $gr(g_f) \subset W^u$. We want the converse

<u>Claim</u>: If $(x,y) \in W^u$, then $y = g_f(x)$.

We simply note that if (x,y) is not on $gr(g_f)$, then since vertical distances are shrunk under f, they are expanded by f^{-1}; if $(x,y) \in W^u$, then its orbit under f^{-1} stays in a bounded ball - but its vertical distance from the invariant set $gr(g_f)$ approaches infinity, a contradiction.

At this point, we have proven the existence of g_f with properties (1) and (4) (the latter follows from (1) and lemma 2, and the fact that $(\tau^{-1} - \varepsilon) > 1$). Our second major step is to show g_f smooth. So we assume $f \in \mathrm{Diff}^K$, and wish to show $g_f \in C^K$.

The idea of the proof is to construct a "tangent bundle" to $gr(g_f)$, which is the same thing as the graph of a derivative of g_f. We formulate the idea of "tangent transform" by defining

$$\Delta f: (E \times E)(r) \longrightarrow E \times E$$

via

$$\Delta f(x,y) = (f(x), Df_x(y)) .$$

If we define $\mathcal{G}^K$ to be the space of C^K maps in $\mathcal{G}$, then we can define, for each $g \in \mathcal{G}^K$, the map

$$\Delta g: E_1(r) \times E_2(r) \longrightarrow E_2(r) \times E_2(r)$$

via

$$\Delta g(x_1, y_1) = (g(x_1), Dg_{x_1}(y_1))$$

and try to define a new graph transform, $\Gamma_{\Delta f}$, acting on the Δg's. However, although Γ_f takes $\mathcal{G}^K$ into $\mathcal{G}^K$, it is by no means a contraction, and what's more Δf is <u>not</u> a priori a Lipschitz perturbation of $\Delta T = T \times T$.

However, at this point, the fiber contraction lemma (lemma 3) will allow us stil to show that the action $\Gamma_{\Delta f}$ of Γ_f on "line bundles" has an attractive fixedpoint.

Formally speaking, if we denote by $\mathcal{H}$ the unit ball in $C^0(E_1(r), L(E_1, E_2))$, realizing then that an element $h \in \mathcal{H}$ is a continuous assignment, to each element of $E_1(r)$, of a non-expanding linear map: $(\forall x_1 \in E_1(r))\ h(x_1): E_1 \longrightarrow E_2$ with $\| h(x_1) \| \leq 1$, then f induces a "graph transform" action

$$\Gamma_{\Delta f} : \mathcal{G} \times \mathcal{H} \longrightarrow \mathcal{G} \times \mathcal{H},$$

defined as follows: Given $(g, h) \in \mathcal{G} \times \mathcal{H}$, we need to define a pair of maps

$$\Gamma_{\Delta f}(g, h) = (\tilde{g}, \tilde{h}) \in \mathcal{G} \times \mathcal{H}.$$

The first factor will simply be

$$\tilde{g} = \Gamma_f\, g .$$

The second factor is to be a continuous map

$$\tilde{h} : E_1(r) \longrightarrow L(E_1, E_2) ,$$

which is defined by its value at a point $x_1 \in E_1(r)$. If we let

$$\xi = (x_1, \Gamma_f g(x_1))$$

then

$$\tilde{h}(x_1) = \Gamma_{Df_\xi}(h(\xi_1)) .$$

It would be tempting to write $\tilde{h}(x_1)$ as $Df_\xi \circ h(x_1) \circ Df_\xi^{-1}$, but this is only true if Df_ξ preserves the splitting. In general it does not, since it is only the derivative of f at a point <u>near</u> the origin; however, we can return to our previous arguments concerning graph transforms, taking note of the fact that we only assumed f was Lipschitz-near T to prove that Γ_f was a contraction of $\mathcal{G}$. Then for ξ sufficiently small, the continuity of Df implies that Df_ξ has a well-defined graph transform Γ_{Df_ξ} which contracts the space of maps $C(E_1, E_2)$ with Lipschitz constant ≤ 1. But since Df_ξ is also linear, it in particular takes lines to lines, and hence (since $\mathcal{H} = \mathcal{G} \cap L(E_1, E_2))$ contracts fibers - provided, again, that ξ is near 0. But by definition,

$$\|\xi\| \leq \|x_1\|$$

so that picking $r > 0$ sufficiently small, $\Gamma_{\Delta f}$ will contract fibers. Thus, by the fiber contraction lemma, $\Gamma_{\Delta f}$ has a unique attractive fixedpoint, $(g_f, h_f) \in \mathcal{G} \times \mathcal{H}$.

We wish to establish that h_f represents a "tangent" to g_f. But it is easy to see that

$$\Delta(\Gamma_f g) = \Gamma_{\Delta f}(\Delta g)$$

whenever g is differentiable. In particular, if we pick an initial function g_o, its iterates

$$\Delta(\Gamma_f^n g_o) = \Gamma_{\Delta f}^n(\Delta g_o)$$

converge to (g_f, h_f); so that if g_o was picked differentiable (for example, the constant zero function) then the tangents to the graph transforms of g_o converge to h_f, so that

$$h_f = D g_f.$$

This shows our assertion for $K = 1$. For $K \geq 2$, we need an induction step. Our induction hypothesis is that the C^K-norm of f is $\| f \|_K \leq M$, and that the theorem holds for $K-1$ (≥ 1)- ie, that $g_{\tilde{f}}$ is C^{K-1} on $E_1(r)$, provided $\tilde{f}$ is C^{K-1} and an ε-perturbation of a linear operator $\tilde{T}$ of skewness τ.

We will show that our g_f is C^K on a smaller ball.

Pick constants ε', r' satisfying

$$\varepsilon < \varepsilon' < \frac{1-\tau}{1+\tau}$$

$$0 < r' \leq r$$

$$Mr' < \varepsilon' - e$$

$$r' \leq 1 .$$

Then the induction step clearly follows from the following:

<u>Claim</u>: $\Delta f'$ is Lipschitz ε'-near $T \times T$, where $f' = f|E(r')$.

<u>Proof</u>

Estimate again:

$$\begin{aligned}
&\| [\ \Delta f' - T \quad T](x,y) - [\ \Delta f' - T \quad T]\ (x', y')\| \\
&\quad \leq \max \{\ \| (f' - T)(x) - (f' - T)(x')\ \| , \\
&\qquad \| (Df'_x - T)(y) - (Df'_{x'} - T)y'\ \|\ \} \\
&\quad \leq \max\{\varepsilon\ \| x-x' \|\ ,\ \| (Df'_x - f'_{x'})(y)\ \| + \| (Df'_{x'} - T)(y-y')\ \|\ \} \\
&\quad \leq \max\{\varepsilon\ \| x-x' \|\ ,\ M\ \| x-x' \|\ r' + \varepsilon\ \| y-y' \|\ \} \\
&\quad \leq \varepsilon' \max\{\ \| x-x' \|\ ,\ \| y-y' \|\ \} \\
&\quad = \varepsilon'\ \| (x,y) - (x' - y')\ \| .
\end{aligned}$$

Thus, by the induction step, $\Delta g_f = (g_f, h_f)$ is C^{K-1}; hence g_f is C^K, and (2) is proven. Finally, to see (3), we note that the fixedpoint of the contraction Γ_f varies continuously with Γ_f, hence with f.

Recall that we assumed $f(0) = 0$. If in fact $f(0) \neq 0$, all our arguments are simply conjugated by a δ-small translation, which changes no Lipschitz constants or norms of maps. We leave it to the reader to check details, that this case in fact also works. //

5.3 STABLE MANIFOLD THEORY FOR A HYPERBOLIC SET

In this section, we will generalize the results of the previous section to a hyperbolic set. For this section, we assume the following situation: M is a (C^∞) Riemannian manifold, U an open subset. $f: U \longrightarrow M$ is a C^K embedding $(K \geq 1)$ and $\Lambda \subset U$ is a compact subset of U, invariant under f (in the sense that $f\Lambda = \Lambda$) and with a hyperbolic structure. For simplicity, we shall usually assume the metric is adapted to Λ.

First, of course, we must formulate the notion of "stable" and "unstable" manifolds. In the case of a fixedpoint, recall that $W^s(p)$ could be characterized as the set of all points whose ω-limit set consists of $\{p\}$, and $W^u(p)$ those with α-limit $\{p\}$. It is clear, however, that in the case of an element p of a hyperbolic set, this would in general be a useless definition, since the set need not even be a subset of the non-wandering set (for instance, our heteroclinic example). However, it is reasonable to look for the set of points whose orbit is asymptotic to that of p, in the sense that $\lim d(f^n(p), f^n(x)) = 0$, where the limit is taken as $n \longrightarrow +\infty$ for $W^s(p)$, and as $n \longrightarrow -\infty$ for $W^u(p)$. It is this notion which we try to formulate more carefully.

Recall that $W^s(p)$ for p fixed could be characterized as $\bigcap_{n \geq 0} f^n[E(r)]$, where $E(r)$ was a neighborhood of p. In the case at hand, given a point $x \in U$ and constant $\beta > 0$, denote the "stable set" of x of "size β" by

$$\Sigma(x, \beta) = \bigcap_{n \geq 0} f^{-n} [N_\beta(f^n x)]$$

where $N_\beta(y)$ denotes the ball of radius β about y. This can be reformulated as

$$\Sigma(x, \beta) = \{y \mid f^n y \text{ is defined and within distance } \beta \text{ of } f^n x\text{—ie, } d(f^n x, f^n y) < \beta, \ (\forall n \geq 0)\}$$

It is not priori clear that staying at bounded distance from the orbit of x is equivalent to being asymptotic to $O(x)$; this will come out of the body of the theorem. We can provisionally adopt the above notion as our definition of "stable set".

Definition: Given constants $\beta \geq \alpha \geq 0$, we say that a submanifold $W \subset U$ is a stable manifold of size (β,α) through $x \in U$ iff

$$W = \Sigma(x, \beta) \cap N_\alpha(x)$$

—ie, the stable manifold is defined if the stable set of size

β, at least within an α-ball about x, is a submanifold.

Another concept which will be useful in our formulation of the stable manifold theorem is that if a continuous family of manifolds.

<u>Definition</u>: A collection $\{W_x\}_{x \in \Lambda}$ of C^k submanifolds (of U) with $x \in W_x$ is called a <u>continuous family of C^k submanifolds</u> iff $(\forall x \in \Lambda)$ there is a neighborhood A of x in Λ, and a continuous map $\varphi : A \longrightarrow C^k(D^n, M)$, (where D^n is the n-dimensional disc) such that for each $x \in A$, φ_x takes D^n diffeomorphically onto a neighborhood of x in W_x.

We can now formulate the stable manifold theorem for hyperbolic sets. Note that whereas the existence of unstable manifolds was easier to deal with in the case of a point, it is the stable manifold we treat in the case of a set. As before, however, the unstable case follows by treating f^{-1}.

<u>Theorem 3</u> (Stable Manifold Theorem for a Hyperbolic Set): Let $\Lambda \subset U \subset M$ be a compact hyperbolic set for C^k embedding $f: U \longrightarrow M$, k a positive integer. Then:

(1) $(\exists\ \beta \geq \alpha > 0)$ such that through each $x \in \Lambda$ there passes a stable manifold W^s_x of size (β, α).

(2) The collection $\{W^s_x\}_{x \in \Lambda}$ of stable manifolds forms a continuous family of C^k submanifolds of U.

(3) There exist constants $K > 0$, $0 < \lambda < 1$ such that for any two points on the same stable manifold, $y, z \in W^s_x$,

$$d(f^n y, f^n z) \leq K\lambda^n d(y, z) \quad (\forall\ n \geq 0)$$

(4) For each $x, y \in \Lambda$,

$$W^s_x \cap W^s_y$$

is an open subset of W^s_x.

(5) W^s_x is tangent at $x \in \Lambda$ to E^s_x, the "stable" component of the hyperbolic structure.
(6) If the metric is adapted, we can take $\alpha = \beta$ in (1) and $K = 1$ in (3).

We could summarize the theorem by saying that for a compact hyperbolic set Λ, local stable manifolds (1) exist, (2) are smooth and vary continuously from point to point of Λ, (3) are contracted by f, (4) overlap in open subsets, and (5) are tangent to the splitting.

Proof

The basic idea of the proof is quite simple, and conforms to our general approach toward such "global" problems. We consider the space

$\mathcal{M}(\Lambda, M)$

of bounded maps of Λ into M. This is a Banach manifold. We locally define a map $\widetilde{F}: \mathcal{M}(\Lambda, U) \longrightarrow \mathcal{M}(\Lambda, M)$ by $\widetilde{F}(h) = f \circ h \circ f^{-1}$.

By lemma 1, this has a hyperbolic fixedpoint at the inclusion $i: \Lambda \hookrightarrow M$. Then by the stable manifold theorem for a point, i has a stable manifold $\mathcal{W}$ for $\widetilde{F}$. This is a submanifold of the space of maps, and the heart of the proof is in showing that the collection of images $W_x = ev_x(\mathcal{W})$ constitutes a family of stable manifolds through Λ.

We shall, however, work somewhat more explicitly in terms of local "coordinates" about $i \in \mathcal{M}$. We recall a few facts from chapter 0. Pick $r > 0$ such that for each $x \in \Lambda$, the exponential map at x maps the ball $M_x(r)$ of radius r in M_x diffeomorphically onto the ball $B_r(x)$ of radius r about x in M, and that furthermore

$$B_r(x) \subset U \cap f^{-1}[U]$$

This is possible because Λ is compact. Then a coordinate patch for $\mathcal{M}$ at i is given by the space $\Gamma_b(V_r)$ of sections (not necessarily continuous) of the r-ball bundle

$$V_r = \{y \in T_\Lambda M \mid \| y \| \leq r\}$$

- that is, the collection of vectorfields defined on Λ and having pointwise norm bounded by r. Our coordinate map φ_i takes this open subset of the Banach space $\Gamma_b(T_\Lambda M)$ onto the open set of maps in $\mathcal{M}$ uniformly within distance r of i, via

$$\varphi_i : \Gamma_b(V_r) \longrightarrow \mathcal{M}(\Lambda, U)$$

$$\varphi_i(\gamma)(x) = \exp_x \gamma(x) .$$

In terms of these coordinates, the local expression for the map $\tilde{F}(h) = f \circ h \circ f^{-1}$ will be denoted

$$F: \Gamma_b(V_r) \longrightarrow \Gamma_b(T_\Lambda M)$$

and given by

$$F(\sigma)_{(x)} = (\exp_x)^{-1} \circ f \circ \exp_{f_x^{-1}} \sigma_{(f_x^{-1})} .$$

We shall find it useful to denote the function composing with σ on the left in the above expression by $\hat{F}$, so that the formula can be rewritten as the functional equation

$$F(\sigma) = \hat{F} \circ \sigma \circ f^{-1} .$$

Diagrammatically, F is given by

$$\begin{array}{ccc} \Lambda & \xrightarrow{F(\sigma)} & U \\ {\scriptstyle f|\Lambda}\downarrow & & \downarrow{\scriptstyle f} \\ \Lambda & \xrightarrow[\sigma]{} & M \end{array}$$

It is fairly standard that F is C^K, with derivative at the zero-section

$DF_0(\sigma) = Tf \circ \sigma \circ f^{-1}|\Lambda$.

The splitting of the tangent space at i, $\Gamma_b = \Gamma_b(T_\Lambda M)$, corresponding to the hyperbolicity of DF_0 is

$\Gamma_b = \Gamma_b({V^s}_r) \oplus \Gamma_b({V^u}_r)$

where clearly ${V^s}_r = V_r \cap E^s$, ${V^u}_r = V_r \cap E^u$. Thus, by the stable manifold theorem for a point, F has a stable manifold $W^s(F) \subset \Gamma_b$, given as the graph of a function

$G\colon \Gamma_b({V^s}_r) \longrightarrow \Gamma_b({V^u}_r)$

which is characterized by

$G(\sigma)$ = the unique section of ${V^u}_r$ such that

$F^n(G(\sigma), \sigma) \in \Gamma_b({V^s}_{s_r}) \qquad (\forall\, n \geq 0)$

That is, given a vectorfield σ mapping into V^s_r , there is a unique vectorfield τ mapping into ${V^u}_r$ such that the sum $\sigma + \tau \in \Gamma_b(V_r)$ stays within distance r of the zero-section under iteration by F, and this sum is the corresponding element of $W^s(F)$.

We note that $W^s(F)$ is a submanifold of the space of sections, so that the union of all images of elements of $W^s(F)$ need not, a priori, have any nice properties. However, the following two lemmas establish the polite behavior of these images, at least on the level of the tangent bundle.

Lemma: There exists a unique fiber map $H: E^s(r) \longrightarrow E^u(r)$, covering the identity id_Λ, such that for each vectorfield $\sigma \in \Gamma_b(V^s{}_r)$, $G(\sigma) = H \circ \sigma$.

Proof

We expand on the ideas of lemma 3, chapter 3. Given a vector $y \in E^s(r)$, we define a section $\sigma_y \in \Gamma_b(V^s{}_r)$ by

$$\sigma_y(x) = \begin{cases} y & y \in M_x \\ 0 & \text{otherwise} \end{cases},$$

We then define the function H by setting

$$H(y) = G(\sigma_y)_{(x)} .$$

To show that H satisfies the required condition, suppose we have any vectorfield σ whose value at x is y:

$$\sigma(x) = y .$$

Then we want to show that

$$G(\sigma)_{(x)} = H(y) .$$

Well, each of the above terms satisfies the inequalities

$$\| \hat{F}^n(G(\sigma)_{(x)}, y) \| \leq r$$
$$\| \hat{F}^n(H(y), y) \| \leq r$$

so that both the vectorfields $(H \circ \sigma) \oplus \sigma$ and $G(\sigma) \oplus \sigma$ remain at bounded distance r from the zero-section under F, and hence by uniqueness are equal.

We note that, as constructed, H is only a set-theoretic map, and need not even be continuous. Let us examine in detail the definition of H: we denote the assignment $y \longrightarrow \sigma_y$

by $\psi : E^s(r) \longrightarrow \Gamma_b(V^s{}_r)$. Then H is given by the composition

$$E^s(r) \xrightarrow{(\psi,\pi)} \Gamma_b(V^s_r) \times \Lambda \xrightarrow{(G,1)} \Gamma_b(V^u_r) \times \Lambda \xrightarrow{ev} E^u(r)$$

All the relevant functions are continuous (in fact smooth), except ev. The evaluation map would be continuous (and linear on fibers) if instead of Γ_b we were considering Γ_o, the continuous sections. But the obstacle to this is that σ_y is not continuous. We remedy the situation by creating a "continuous" version of ψ . This construction is a general topological one, and is the "continuous" version alluded to in lemma 3, chapter 3.

Lemma: Suppose $E \xrightarrow{p} B$ is a (continuous) vector bundle over a normal topological space, and $y \in E$.
Then there exist a neighborhood U of y in E and a neighborhood V of py in V, with $pU \subset V$, and a map

$$\varphi : U \longrightarrow \Gamma_o(E)$$

such that
(1) φ is continuous and linear (where defined) on fibers in U
(2) The support of $\varphi(x) \in \Gamma_o(E)$ is contained in V for each $x \in U$.
(3) $\varphi(x)$ takes the value x at px.
Furthermore, if $\| y \| < r$, we can pick φ so that $\| \varphi(x) \| \leq r \quad (\forall x \in U)$.

Thus, φ is a continuous assignment to vectors in U of continuous vectorfields which equal the point when they can, and this assignment can give vectorfields uniformly as close to the zero-section as the vectors are to the origin.

Proof of Lemma

The lemma takes less time to prove than to state. We pick V a trivializing neighborhood (of py) for p, so that locally, p is just the projection

$$\begin{array}{c} V \times F \\ \downarrow p \\ V \end{array}$$

where F is the fibering space. We then pick W a neighborhood of py whose closure is contained in V and $\psi : E \longrightarrow \mathbb{R}$ a Urysohn function for $\overline{W}$ — ie, $\psi = 0$ outside V and 1 inside W. Finally, we define

$$\varphi(x)_{(\xi)} = (\xi, \psi(\xi)x_2)$$

where the right hand is zero unless $x \in V \times F$ and $\xi \in V$, in which case $x = (x_1, x_2)$. This clearly satisfies (1)-(3).

Now, we can deduce

Corollary: H is continuous and C^K on fibers.

Proof

We can re-express H as

$$U \xrightarrow{(d,\pi)} \Gamma_o(V_r^s)\times\Lambda \xrightarrow{(\tilde{G},1)} \Gamma_o(V_r^u)\times\Lambda \xrightarrow{ev} E^u(r)$$

where φ is as above and $\tilde{G}$ is $G|\Gamma_o$. Then by uniqueness, this H agrees with the old one, and here all functions are continuous (including ev) and φ is linear, $\tilde{G}$ is C^K, and ev is C^∞ on fibers, so that H is C^K on fibers.

Now, we let $\mathcal{W}$ denote the graph of H. This is a subset of V_r, namely

$$gr(H) = \{(Hy, y) \in E^u(r) \times E^s(r) \mid y \in E^s(r)\}$$

We then project down to U in two steps: first, given $x \in \Lambda$, define

$$\mathcal{W}_x = \mathcal{W} \cap M_x(r)$$

so that $\mathcal{W}_x$ is a C^K submanifold of $M_x(r)$, then let

$$W^s_x = \exp_x(\mathcal{W}_x \sim \partial\mathcal{W}_x)$$

which in turn (since $\exp_x$ is C^∞) will be a C^K submanifold of U. Since H is continuous, the W^s_x clearly form a continuous family of C^K submanifolds. That W^s_x is contracted by f follows from the fact that $W^s(F)$ is contracted by F; the open overlap follows from the definitions, as does the fact that W^s_x is a local stable manifold through x, and the fact that W^s_x is tangent to E^s_x. Thus we are done with the stable manifold theorem for a hyperbolic set.

Note in particular that the existence of stable and unstable manifolds has nothing to do with whether Λ consists of non-wandering points. All that is required is compactness and a hyperbolic structure on Λ. //

CHAPTER 6

THE Ω-STABILITY THEOREM

In this chapter, we carry further our analysis of hyperbolic sets from chapter 5 in the case when the hyperbolic set in question is the whole non-wandering set, $\Omega(f)$. This analysis was outlined by Smale in [3], and more details were given in Smale [5], with a final important detail filled in by Hirsch, Palis, Pugh and Shub [1]. The main result is the Ω-stability theorem, giving sufficient conditions for a diffeomorphism of a compact manifold to be Ω-stable. The arguments in some ways extend those of Palis [2] and Palis and Smale [1] on structural stability of Morse-Smale diffeomorphisms.

The analysis proceeds in four large steps. First, the spectral decomposition theorem establishes the analogy with Morse-Smale systems (under appropriate assumptions) by decomposing the non-wandering set into a finite disjoint union of closed, invariant, "indecomposable" sets. Then, some facts about the behavior of points near one of these "pieces" of are established. Third, local Ω-explosions are eliminated. Finally, the filtration construction globalizes the no-explosion result.

Our major assumption, first formulated by Smale [3], is

<u>Axiom A</u>: (a) $\Omega(f)$ has a hyperbolic structure.

(b) periodic points are dense in $\Omega(f)$.

We henceforth restrict ourselves to compact manifolds. The "general density theorem" of Pugh [2] states that property (b) is C^1-generic. Property (a) is definitely not dense, as several recent counterexamples (which we shall discuss briefly in the final chapter) show. It is a standing conjecture that (a) implies (b), and similarly a standing conjecture that axiom A is a necessary condition for Ω-stability. Some results toward the second conjecture have recently been found by Franks ([3] and [4]).

6.1 SPECTRAL DECOMPOSITION

We establish in this section that a diffeomorphism satisfying axiom A behaves vaguely like a Morse-Smale system, in the sense that Ω has a finite number of "indecomposable" pieces. To clarify what this means, we recall a definition from topological dynamics.

Definition: A homeomorphism h of a topological space X is *topologically transitive* if some point of x has a dense orbit.

If X is an invariant subspace of some larger space on which h is also defined, we can by abuse of notation refer to X as a "transitive" subset.

We are concerned with decomposing $\Omega(f)$ into closed, disjoint invariant subsets; it therefore seems entirely natural to call a set indecomposable if it cannot be so decomposed.

Clearly, any transitive set is indecomposable. On the other hand, the converse need not be true, since if $\omega(x) \cap \omega(y) \neq \emptyset$ and $x \notin \overline{O(y)}$, then $\overline{O(x) \cup O(y)}$ is indecomposable but not transitive.

It is also instructive to point out the distinction between transitive (or even indecomposable) and minimal sets.

Again, every minimal set is transitive (hence indecomposable), but an infinite minimal set cannot, for example, contain periodic points, while the shift automorphism (Ω in the horseshoe) is transitive, with periodic points also dense.

We now have all the concepts we need to state

Theorem 1 (Spectral Decomposition Theorem): If M is compact and $f:M \longrightarrow M$ is a diffeomorphism satisfying axiom A, then there is a unique decomposition of $\Omega(f)$

$$\Omega(f) = \Omega_1 \cup \dots \cup \Omega_K$$

into a finite union of disjoint, closed, invariant and transitive sets.

The Ω_i in the above decomposition are usually referred to as basic sets.

Corollary: $M = \bigcup_{i=1}^{K} \tilde{W}^s(\Omega_i)$, where $\tilde{W}^s(\Omega_i) = \{x \in M | f^m(x) \longrightarrow \Omega_i \text{ as } m \longrightarrow +\infty\}$.

Note that, a priori, the stable manifold of a set need not be the union of stable manifolds through its points. We shall, however, establish this fact in a later section for Ω_i.

The proof of both theorem 1 and the corollary will be given after a series of technical results.

Smale [3] gives the following example (which he attributes to Jacques Tits) to motivate the title "spectral decomposition" for theorem 1. Let $u: V \longrightarrow V$ be a linear transformation of complex n-dimensional space, and assume that u has determinant 1 (ie, $u \in SL(n, \mathbb{C})$) that the eigenvalues, λ_i, of u all have distinct absolute values, and none has absolute value one. Then u induces a diffeomorphism u_0 of (n-1)-dimensional complex projective space

$u_0 : \mathbb{C}P^{n-1} \longrightarrow \mathbb{C}P^{n-1}$,

with $\Omega(u_0) = \{e_i\}$, where the e_i are the projections of the eigenvectors, and the linear "spectral decomposition" of u projects to the dynamic "spectral decomposition" of $\Omega(u_0)$.

Our first technical lemma concerns the global behavior of stable and unstable manifolds through (hyperbolic) periodic points. This is a weak version of the "λ-lemma" proven by Palis in [2], and is a global version of prop. 2 (b), chapter 5. Recall that the (global) stable manifold $W^s(p)$ of a fixedpoint p is obtained by iterating backward the local stable manifold given by either Hartman's theorem or the stable manifold theorem.

Lemma 1: (a) If p is periodic for f and U is an open set intersecting $W^s(p)$, then the closure of the positive orbit of U contains the unstable manifold of p:

$$U \cap W^s(p) \neq \emptyset \implies \text{clos}[\bigcup_{m > 0} f^m[u]] \supset W^u(p) .$$

(b) If furthermore, q is another periodic point whose stable manifold has a transverse intersection with $W^u(p)$, then the positive orbit of U intersects $W^s(q)$:

$$W^u(p) \pitchfork W^s(q) \neq \emptyset \implies W^s(q) \cap \bigcup_{m>0} f^m[U] \neq \emptyset .$$

Proof

By considering an appropriate iterate, we can assume p and q fixed by f.

(a): $W^u(p)$ is expanded by f, so it will suffice to show that the closure of the positive orbit of U contains a neighborhood of p in $W^u(p)$. Since $U \cap W^s(p) \neq \emptyset$, some iterate will take an open subset of U, intersecting $W^s(p)$, into a neighborhood of p on which f is conjugate to $(Df)_p$, by Hartman's theorem. But then we are reduced to the case of prop. 2(b), chapter 5, and we are done.

(b) This follows fairly easily from a topological argument — $W^s(q)$ contains a transverse section to $W^u(p)$, and since U "piles up" on $W^u(p)$, they must intersect. //

The second lemma is a kind of analogue of the fact that homoclinic points are non-wandering (and in fact part of a shift, by theorem 2, chapter 4); we show here that certain heteroclinic points are also necessarily in Ω.

Lemma 2 ("Cloud Lemma"): If p_i $(i=0,\dots,n)$ are periodic points with $p_0 = p_n$, and if x_i is a point of transversal intersection of $W^u(p_i)$ with $W^s(p_{i+1})$, $(i=0, \dots, n-1)$, then each x_i is non-wandering.

Proof

Take a neighborhood U_i in M of each x_i; by lemma 1(b) and induction,

$$W^s(p_j) \cap \bigcup_{m>0} f^m[U_i] \neq \emptyset$$

for each j; in particular, then, by lemma 1(a),

$$\mathrm{clos}[\bigcup_{m>0} f^m(U_i)] \supset W^u(p_i) ,$$

and it immediately follows that $x_i \in \Omega(f)$. //

These two lemmas allow us to prove a relatively important result about the structure of Ω, called the "product neighborhood structure", or

Proposition 1 (Canonical Coordinates): If f satisfies axiom A and $x \in \Omega(f)$, then there exists a constant $\varepsilon > 0$, uniform for Ω, and a map

$$I_x : V \longrightarrow M$$

where V is a neighborhood of $x \times x$ in

$$\check{W}^u(x) \times \check{W}^s(x) \quad (\check{W}^s(x) = W^s(x) \cap \Omega)$$

such that

1) I_x is a homeomorphism onto a neighborhood of x in Ω
2) $I_x | \check{W}^u(x,\varepsilon) \times \{x\} = J^u_x$, the inclusion of $\check{W}^u(x,\varepsilon)$ in M
3) $I_x | \{x\} \times \check{W}^s(x,\varepsilon) = J^s_x$
4) $I_x(p,q)$ is defined by

$$I_x(p,q) = W^s(p,\varepsilon) \cap W^u(q,\varepsilon)$$

Proof

If I_x is well-defined by (4), then (2) and (3) follow automatically. From the stable manifold theorem, $W^s(p,\varepsilon)$ and $W^u(q,\varepsilon)$ are very nearly equal to $W^s(x,\varepsilon)$, $W^u(x,\varepsilon)$ (resp.); in particular, by stability of transversal intersection (theorem 1(b), section 0.4) they must intersect in a unique point. Thus, I_x is well-defined. By the cloud lemma, its image is in

Ω. It is onto a neighborhood of x in Ω by a modification of the above transversality argument. It is injective because $W^s(x) \cap W^s(y)$ is open in $W^s(x)$, so that if $W^s(p,\varepsilon) \cap W^u(q,\varepsilon) = W^s(\tilde{p},\varepsilon) \cap W^u(\tilde{q},\varepsilon)$ then $W^s(p,\varepsilon) \cap W^s(\tilde{p},\varepsilon) \neq \emptyset \Longrightarrow W^s(p,\varepsilon) = W^s(\tilde{p},\varepsilon)$, and ε could be chosen (uniformly) small so that any two invariant submanifolds intersect only once (ie, since p, $\tilde{p} \in W^s(p,\varepsilon) \cap W^u(x,\varepsilon)$, $p = \tilde{p}$). This completes the proposition. Note that it shows that if we restrict ourselves to a small enough neighborhood N of x, the stable and unstable manifolds of points of $\Omega \cap N$ form a "grid" on N, with $\Omega \cap N$ consisting of the "lattice" points. //

The next result gives us a means of defining the "pieces" so as to be transitive.

Lemma 3 (Local Transitivity): If N is a neighborhood of $x \in \Omega(f)$ as in prop. 1, and U a nonempty subset of N, then

$$\bigcup_{m \geq 0} f^m[U] \quad \text{and} \quad \bigcup_{m \geq 0} f^{-m}[U]$$

are each dense in N.

Proof

Since periodic points are dense, we can pick $q \in U \cap \mathrm{per}(f)$ and $p \in N \cap \mathrm{per}(f)$, and it suffices to show $p \in \mathrm{clos} \bigcup_{m \geq 0} f^m[U]$. But by the definition of N, $W^u(p)$ and $W^s(q)$ intersect transversally at a unique point, say x_i; similarly, $W^u_\varepsilon(q)$ and $W^s_\varepsilon(p)$ intersect transversally, say at x'. By the cloud lemma, $x, x' \in \Omega(f)$, and

$x \in \bigcup_{m \geq 0} f^{-m}[U]$ by lemma 1(b). Thus,

$p \in \text{clos} \bigcup_{m \geq 0} f^{-m}[U]$, by lemma 1(a), and $O_+[U]$ is dense in N. //

Proof of Theorem 1

The spectral decomposition is now easy to define: given $x \in \Omega(f)$ pick $N(x)$ a neighborhood of x as in proposition 1 and define

$$\Omega_x = \text{clos} \bigcup_{m \in \mathbb{Z}} f^m[N(x)] .$$

Then clearly each Ω_x is closed, invariant, and transitive. To see that any two are either disjoint or identical, note that Ω_x contains a neighborhood of x, so that if $\Omega_x \cap \Omega_y \neq \emptyset$, then $x \in \text{clos}\, O(N(y))$, hence for some $n \in \mathbb{Z}$,

$$U = f^n[N(y)] \cap N(x) \neq \emptyset$$

and by lemma 3, we can argue that

$$\Omega_x = \text{clos}\, O[U] = \Omega_y .$$

Thus, distinct Ω_x's are disjoint. Similarly, by compactness, the fact that each Ω_x contains the neighborhood $N(x)$ of x implies (finite covering) that there are only finitely many distinct Ω_x's. The transitivity in the sense of a dense point-orbit follows from the fact that any open subset in Ω_i has a dense orbit, using a standard argument which we will separate out as a lemma, since it has other applications. Modulo the lemma, we are done. //

Lemma 4: If X is a separable topological space and

$h: X \longrightarrow X$ a homeomorphism such that any nonempty open set has dense orbit, then the set of points with dense orbits is residual.

Proof

Let $\{V_i\}_{i \in \mathbb{Z}_+}$ be a countable basis for the topology. Then the orbit of each V_i,

$$O[V_i] = \bigcup_{n \in \mathbb{Z}} f^n[V_i]$$

is a dense, open set, so the intersection over all the V_i

$$U = \bigcap_{i=1}^{\infty} O[V_i]$$

is residual. On the other hand, if $x \in U$, then x belongs to the orbit of each V_i; in other words, each V_i contains some iterate of x, and hence x has dense orbit. //

Clearly, since in the case at hand X is a compact subset of a manifold, it is a Baire space; therefore the set of points with dense orbits, being residual, is non-empty.

We now proceed to the corollary of theorem 1, whose proof is not entirely trivial.

Proof of Corollary

Since M is compact, it is clear that $\omega(x) \neq \emptyset$ for each $x \in M$. Obviously $\omega(x) \subset \Omega(f)$, but we wish to show that each $\omega(x)$ is contained in a single Ω_i.

We pick two families of disjoint, open sets, satisfying

$$V_i \supset U_i \supset \Omega_i$$

and with

$$f^{-1}[U_i] \cup f[U_i] \subset V_i$$

Given V_i, we can clearly pick U_i a neighborhood of the invariant set Ω_i small enough so that the second condition holds.

Now suppose we have $x \in M$ and $K \neq \ell$ such that

$$\omega(x) \cap \Omega_K \neq \emptyset$$
$$\omega(x) \cap \Omega_\ell \neq \emptyset .$$

Then we can pick two sequences of positive integers, $\{m_i\}$ and $\{\ell_i\}$, increasing in the sense that

$$m_j < m_j + \ell_j < m_{j+1}$$

with

$$f^{m_j}(x) \in U_K$$
$$f^{m_j+\ell_j}(x) \in U_\ell .$$

This is because by assumption the forward iterates of x cluster around both Ω_K and Ω_ℓ, so must be frequently in both U_K and U_ℓ. But by assumption, the first iterates of $f^{m_j}(x)$ and $f^{m_j+\ell_j}(x)$ stay in V_K (resp., V_ℓ), so that in particular, for each j, there is an iterate n_j, with

$$m_j < n_j < m_j + \ell_j$$

such that

$$f^{n_j}(x) \notin U_K \cup U_\ell .$$

In fact, by an obvious extension of the argument in case x also clusters around other pieces Ω_i, we can pick the above n_j so that

$f^{n_j}(x) \notin \bigcup_i U_i$.

But then any cluster point of the seauence $f^{n_j}(x)$ cannot lie in Ω , which is an obvious contradiction — so $K = \emptyset$. //

Thus, by the corollary, we see that M decomposes into a disjoint union of stable manifolds of the Ω_i; similarly, we can reverse the map to get $M = \bigcup_i \tilde{W}^u(r_i)$.

6.2 LOCAL STRUCTURE OF Ω_i

Our goal here is to show that the stable manifold of each Ω_i is the union of stable manifolds of points in Ω_i. In other words, any point that is eventually near Ω_i is actually asymptotic to an element of Ω_i.

Recall that we defined

$$\tilde{W}^s(\Omega_i) = \{x \mid f^n x \longrightarrow \Omega_i \quad \text{as} \quad n \longrightarrow +\infty\}$$

$$W^s_\varepsilon(\Omega_i) = \bigcup_{x \in \Omega_i} W^s_\varepsilon(\Omega_i) .$$

So that our statement will become

$$\tilde{W}^s(\Omega_i) = W^s(\Omega_i) .$$

Note that the local product structure (proposition 1) tells us that $\Omega_i = W^s_\varepsilon(\Omega_i) \cap W^u_\varepsilon(\Omega_i)$ for small enough ε, and since f contracts $W^s_\varepsilon(\Omega_i)$, we can also say that

$$\bigcap_{n \geq 0} f^n[W^s_\varepsilon(\Omega_i)] = \Omega_i \qquad (1).$$

These identities, and in fact the main theorem of this section follow from just the local product structure, and throughout this section we will be dealing simply with a compact hyperbolic set Λ with local product structure ($W^s_\varepsilon(\Lambda) \cap W^u_\varepsilon(\Lambda) = \Lambda$). We are following §§3-5 of Hirsch, Palis, Pugh and Shub [1].

To show that the assumption of local product structure cannot be dropped, we note (following R. Bowen) that it suffices to find a compact invariant subset Λ of A^∞_∞, the shift automorphism on two symbols (section 4.1) with periodic points dense in Λ, but with points $c \notin \Lambda$ asymptotic to the set and not to any points in the set. Such a Λ is given by

$\Lambda = \{x \mid$ any finite maximal block of zeroes has even length$\}$

and a c is given by

$\{..1000101.1010001 ...\} = c$

where the bisequence is gotten by tacking on successive odd maximal blocks of zeroes (of increasing length) on each side. Bowen's example is given in fuller detail in Hirsch, Palis, Pugh and Shub [1].

The following notion was first formulated (for fixedpoints) by Palis in [2]:

<u>Definition:</u> Λ a compact hyperbolic set, then a <u>fundamental domain</u>

1) for $W^s_\varepsilon(\Lambda)$ is a compact set

$D \subset W^s_\varepsilon(\Lambda)$

such that

$W^s_\varepsilon(\Lambda) \sim \Lambda \subset O_+[D]$

2) for $W^u_\varepsilon(\Lambda)$ is a compact set

$D \subset W^u_\varepsilon(\Lambda)$

such that

$W^u_\varepsilon(\Lambda) \sim \Lambda \subset O_-[D]$.

(In other words, a fundamental domain for $W^u_\varepsilon(\Lambda)$ under f is a fundamental domain for $W^s_\varepsilon(\Lambda)$ under f^{-1}).

The domain is proper if it is disjoint from Λ .

The existence of fundamental domains for any hyperbolic set is established by

Lemma 5: (1) $D^s = \text{clos}[W^s_\varepsilon(\Lambda) \sim f[W^s_\varepsilon(\Lambda)]]$ is a fundamental domain for $W^s_\varepsilon(\Lambda)$.

(2) $D^u = \text{clos}[W^u_\varepsilon(\Lambda) \sim f^{-1}[W^u_\varepsilon(\Lambda)]]$ is a fundamental domain for $W^u_\varepsilon(\Lambda)$.

Proof

The proof is very simple. For D^s ,

$$O_+[D^s] \supset W^s_\varepsilon(\Lambda) \sim \bigcap_{n \geq 0} f^n[W^s_\varepsilon(\Lambda)]$$

$$= W^s_\varepsilon(\Lambda) \sim \Lambda \qquad \text{by (1) (p. 199)}$$

The argument for D^u is obtained as before by reversing f. //

In the case at hand, we shall need proper domains. Propriety is established by

Lemma 6: If Λ has local product structure, then D^s and D^u are proper.

Proof

We can certainly find a constant $\delta > 0$ such that $0 < \delta < \varepsilon$ and

$f[W^s_\varepsilon] \supset W^s_\delta$.

We claim W^s_δ is a neighborhood of Λ in W^s_ε.

Otherwise, we could pick a sequence of points $x_n \in [W^s_\varepsilon(p_n) \sim W^s_\delta(\Lambda)]$ converging to $p \in \Lambda$. But then (if ε is small enough) $W^s_{2\varepsilon}(p_n)$ and $W^u_{2\varepsilon}(p)$ has a unique point of intersection, q_n, which can (since $W^s_\varepsilon(x_n) \subset W^s_{2\varepsilon}(p_n)$) be regarded as a "projection" of x_n along $W^s(x_n)$ into $W^u_{2\varepsilon}(p)$. By local product structure, $q_n \in \Lambda$, but also, since $x_n \longrightarrow p$, $q_n \longrightarrow p$ and therefore $d(x_n, q_n) \longrightarrow 0$. But then for large n, $x_n \in W^s(q_n)$ (since clearly $W^s(x_n) = W^s(q_n)$) contradicting the assumption on x_n.

Now, since $f[W^s_\varepsilon(\Lambda)]$ is a neighborhood of Λ in $W^s_\varepsilon(\Lambda)$, it must follow that Λ does not intersect the interior or the boundary of $W^s_\varepsilon \sim f[W^s_\varepsilon]$, so $D^s \cap \Lambda = \emptyset$.

The argument for D^u follows as before. //

The next step in the argument is to construct a weak extension of the family $\{W^s_\varepsilon(x)\}_{x \in \Lambda}$ to a neighborhood of in M. To be precise about the kind of extension we require, we restate the notion of a continuous family of submanifolds from p. 181:

<u>Definition</u>: A <u>(continuous)</u> <u>n-disc</u> <u>family</u> through a set $X \subset M$ is a collection $\{D^n_x\}_{x \in X}$ of n-manifolds with $x \in D^n_x$, with D^n_x given as the image of an embedding $\varphi_x : D^n \longrightarrow M$ of the n-disc and with the assignment

$$\varphi : X \longrightarrow \mathrm{Emb}\,(D^n, M)$$

a continuous map.

By unstable manifold theory, $\{W^u(x)\}_{x\in\Lambda}$ is a continuous n-disc family through Λ ; we wish to extend this to a u-disc family $\{W^u(y)\}_{y\in N}$ through a neighborhood N of Λ, satisfying the "overflowing-invariant" condition

$$\overline{W}^u_\delta(fx) \subset f[\overline{W}^u_\delta(x)]$$

whenever it makes sense (ie, when $x \in N \cap f^{-1}N$). Formally stated, this becomes

Proposition 2: If Λ is a compact hyperbolic set, then there exists a neighborhood N of Λ in M and a continuous disc family $\overline{W}^u_\delta$ through N satisfying

(1) $\overline{W}^u_\delta = W^u_\delta$ at Λ

(2) if $x \in N \cap f^{-1}N$, then

$$\overline{W}^u_\delta(fx) \supset f[\overline{W}^u_\delta(x)] .$$

Moreover, the union

$$\bigcup_{y\in W^s_\varepsilon(p)} \overline{W}^u_\delta(y)$$

of discs through the local stable manifold of $p \in \Lambda$ is a neighborhood of p in M.

Proof

We will simply outline the proof, since the arguments closely follow those of 5.2 and 5.3 .

We begin by extending the splitting $E^s \oplus E^u$ over Λ to a (not necessarily invariant) splitting $\overline{E}^s \oplus \overline{E}^u$ over N. This is done by regarding E^s and E^u as sections over Λ of the grassmannian bundle and using a simple extension theorem on sections of fiber bundles (details in Hirsch, Palis, Pugh, Shub [19]). We also construct a "connector" - ie, a class of isomorphisms

$\theta_{(x,x')} : T_xM \longrightarrow T_{x'}M$

defined for (x, x') in some neighborhood of the diagonal in $M \times M$ and satisfying

(1) $\theta_{(x,x)} = \text{id}$
(2) θ preserves the (extended) splitting over N.

This gives us a way of expressing

$(Tf)_x : T_xM \longrightarrow T_{fx}M$

by a matrix with respect to the splitting $\overline{E}^s \oplus \overline{E}^u$

$$\begin{pmatrix} A_x & B_x \\ C_x & K_x \end{pmatrix}.$$

Now, the smooth map $\exp_x^{-1} \circ f \circ \exp_x$ defined on a neighborhood of the zero-section in T_xM is a Lipschitz perturbation of the operator $T: T_xM \longrightarrow T_{fx}M$ given by

$$\begin{pmatrix} A_x & 0 \\ 0 & K_x \end{pmatrix}.$$

for x near Λ.

But then we simply re-interpret the graph-transform argument of 5.2, to get a graph transform defined on maps $\overline{E}_x^u \longrightarrow \overline{E}_x^s$ of Lipschitz constant ≤ 1 and taking values among maps $\overline{E}_{fx}^u \longrightarrow \overline{E}_{fx}^s$. The "unstable" manifold of this new operator will, as before, project to our required n-disc family. The details are left to the reader's imagination, though more of them can be found in Hirsch, Palis, Pugh,

Shub [1]. //

To state the next result we define

<u>Definition</u>: A <u>fundamental neighborhood</u> for W^s_ε is a compact neighborhood of a fundamental domain.

Then the final key to the main result of this section is

<u>Proposition 3</u>: If V is a fundamental neighborhood for $W^s_\varepsilon(\Lambda)$, then

$$W^u_\varepsilon(\Lambda) \cup 0_+[V]$$

is a neighborhood of Λ in M.

<u>Proof</u>

Let D be the fundamental domain for W^s_ε contained in V. Pick a u-disc family $\tilde{W}^u$ on D such that $\tilde{W}^u(y) \subset V$ for all $y \in D$, and with $0 < \delta \leq \varepsilon$. Then this inclusion implies that

$$W^u_\varepsilon \cup 0_+[V] \supset W^u_\varepsilon \cup \bigcup_{y \in D} \tilde{W}^u_\delta(0_+ y) \tag{1}$$

since

$$f^n[V] \supset f^n[\tilde{W}^u(y)] \supset \tilde{W}^u(f^n y) .$$

Also, since D is a fundamental doamin, $W^s_\varepsilon(\Lambda) = \Lambda \cup \bigcup_{y \in D}(0_+ y)$ so that

$$W^u_\varepsilon(\Lambda) \cup \bigcup_{y \in D} \tilde{W}^u_\delta(0_+ y) \supset \bigcup_{y \in W^s_\varepsilon} \tilde{W}^u_\delta(y). \tag{2}$$

Now for any $p \in \Lambda$, certainly

$$\bigcup_{y \in W^s_\varepsilon} \tilde{W}^u_\delta(y) \supset \bigcup_{y \in W^s_\varepsilon(p)} W^u_\delta(y) \tag{3}$$

and it is easy to show by a homotopy argument that this last union is a neighborhood of p in M.

Combining (1), (2) and (3), we get the desired //.

Having gone through all this preliminary work, we can state the main result of this section as a

<u>Corollary</u>: If $W^u_\varepsilon(\Lambda)$ has a proper fundamental domain D, then Λ has a neighborhood U such that for any $x \in M$, if $O_+(x) \subset V$ then $x \in W^s_\varepsilon(p)$ for some $p \in \Lambda$.

<u>Proof</u>

Let V be a compact neighborhood of D disjoint from Λ. Set

$U = (W^s_\varepsilon(\Lambda) \cup O_-[V]) \sim V$.

By proposition 3 applied to f^{-1}, U is a neighborhood of Λ.

But if $f^n(x) \in U$ for each positive n, then $x \notin O_-[V]$, since otherwise we would have (for some $n \geq 0$)

$f^n x \in V$

and $O_+(x) \subset U \subset M \sim V$. Thus, $x \in W^s_\varepsilon(\Lambda)$. //

Now, it is clear that we have the chain of inclusions

$W^s_\varepsilon(\Lambda) \supset \tilde{W}^s_\varepsilon(\Lambda) \supset \{x \mid O_+(x) \in U\} \subset W^s_\varepsilon(\Lambda)$.

6.3 Ω-EXPLOSIONS

It might seem, at first glance, that the "stability" theorem (theorem 1) of chapter 5 should immediately give us an Ω-stability theorem: if the non-wandering set $\Omega(f)$ has a hyperbolic structure and g is a C^1 perturbation of f, then g has a compact hyperbolic set Λ_g near $\Omega(f)$ and conjugate to $\Omega(f)$. Presumably points that wandered before won't all of a sudden stop wandering under a small perturbation, so we should be done.

The trouble with this reasoning is the "presumably" step; in fact axiom A alone does not suffice to give Ω-stability. We begin this section with an example, due to Palis [3], which shows what can go wrong — the phenomenon of "Ω-explosions". The second part of this section will be devoted to the proof (Hirsch-Pugh [1], §7) that Ω-explosions are an essentially global phenomenon. The next section will then show how, under appropriate global assumptions, such phenomena can be avoided.

We begin with Palis' example. Actually, Palis [3] has a general theorem, but we give a specific example to show how the argument works. Our example is two-dimensional and has six fixedpoints — two each of sources, saddles and sinks. Certain stable manifolds, however, coincide with certain other unstable manifolds. Consider the diagram

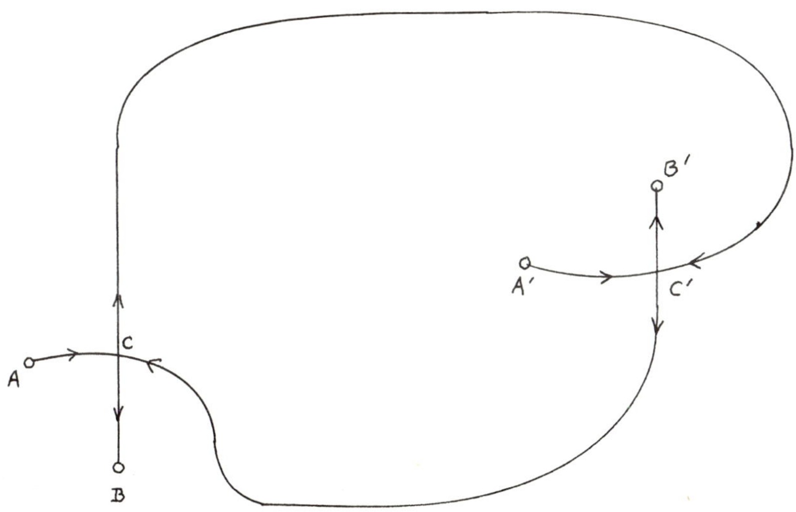

Figure 1a: Preparation for an explosion.

Figure 1b: The old Ω.

Here, A, A' are sources, B, B' are sinks, and C, C' are saddles with stable and unstable directions indicated by the arrows. The crucial fact is the cycle of relations

$$W^s(C) \sim \{C\} = W^u(C') \sim \{C'\}$$

$$W^u(C) \sim \{C\} = W^s(C') \sim \{C'\}$$

Such a diagram can easily be realized in practice, for eaample by putting the whole diagram on the "front" side of a 2-sphere and gluing a similar diagram on the backside. The reader can check that each point on $W^s(C) \sim \{C\}$ or $W^s(C') \sim \{C'\}$ wanders.

Now, we pick a point $D \in W^s(C) \sim \{C\}$ and $D' \in W^s(C') \sim \{C'\}$ and perturb the map slightly on a ball about each, so that $g(W^s_f(C) \cap D)$ has a point of transverse intersection with $W^s_f(C)$. (Similarly for D')

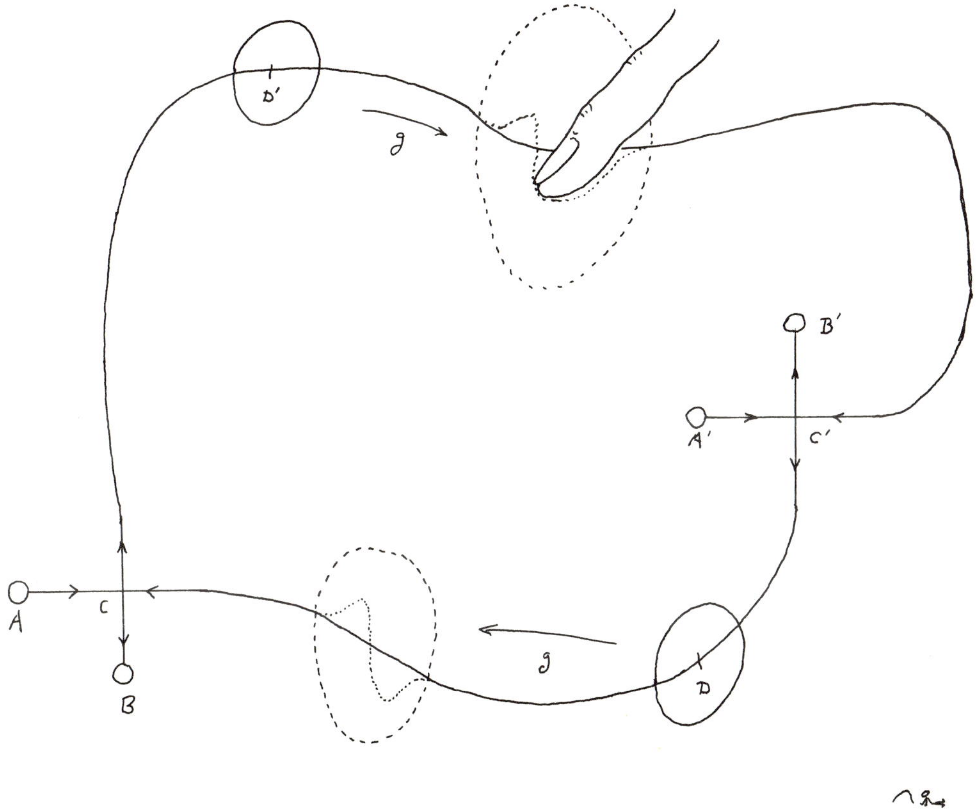

Figure 2a Pushing the plunger

Then we claim the new point of transverse intersection will be non-wandering: note that now $W^u(A') \pitchfork_D W^s(C)$, so that a neighborhood will "pile up" on $W^u(C)$, and in doing so will become transverse to $W^s(C')$, so will "pile up" on $W^u(C')$, and hence U will eventually cut across its own bottom. Under the old map, however, U fell into the sinks before it could ever reach itself again. Thus we have

Figure 2b: An explosion

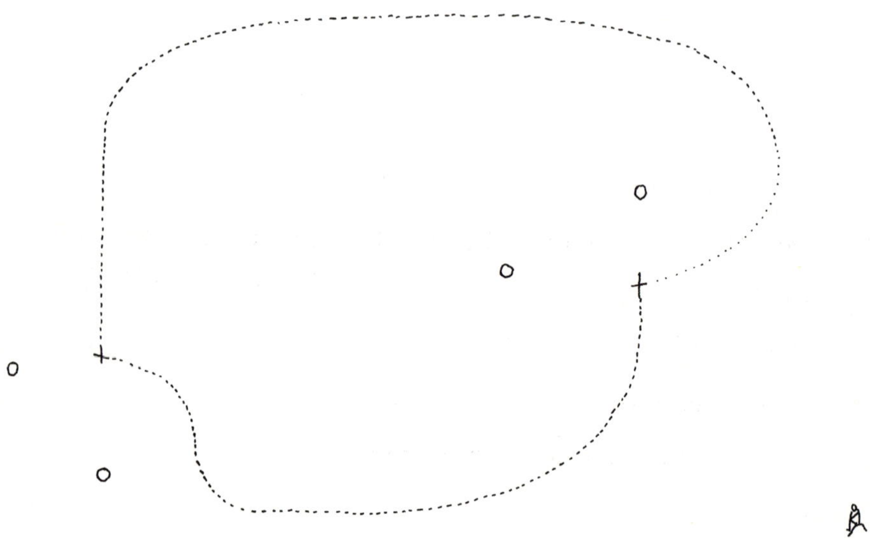

Figure 2c: The new Ω .

Thus, by a perturbation far away from $\Omega(f)$, we can create new non-wandering points arbitrarily near Λ_g. On the other hand, this is a purely global phenomenon, in the sense that the points themselves have orbits that wander quite far from Λ_g. We shall proceed to show this "local" fact, and then in the next section will see how to avoid the global problems.

The rest of this section is taken more or less from § 7 of Hirsch and Pugh [1]. We begin by stating without proof a technical theorem which seems relatively straightforward.

Lemma 7 (Hirsch-Pugh [1], Lemma 4.8): Given $0 < \tau < 1$, there exists $\varepsilon > 0$ such that: if $T: E_1 \times E_2 \longrightarrow E_1 \times E_2$ is a linear transformation given by the matrix

$$\begin{pmatrix} A & B \\ C & D \end{pmatrix} \quad \text{with:}$$

(1) A, D invertible.
(2) $\max\{\|A^{-1}\|, \|D\|\} < \tau + \varepsilon$
(3) $\max\{\|B\|, \|C\|\} < \varepsilon$
then T is hyperbolic.

This seems like a generalization of corollary 1, p. 140 of these notes. The full proof is given by Hirsch and Pugh.

The situation we shall be studying in the rest of this section, then, will be as follows: f will be a C^K embedding of an open subset U of manifold M into M; Λ will be a compact subset of U, with $f\Lambda = \Lambda$ and possessing a hyperbolic structure for f of skewness τ with respect to an adapted metric.

Our goal is to show that, possibly for a smaller U, any sufficiently small perturbation g of f will have exactly the same invariant subsets of U (up to conjugacy), and these will be hyperbolic. We prove this last statement first, formulated as

Proposition 4 (Theorem 7.1, Hirsch-Pugh [1]): Given $\varepsilon > 0$ there exists a neighborhood V of Λ in U and a C^K neighborhood $\mathcal{N}$ of f such that any subset of V invariant under some $g \in \mathcal{N}$ has hyperbolic structure (for g) of skewness σ, with

$|\tau - \sigma| < \varepsilon$.

Proof

We begin by extending the splitting over Λ to a splitting over a neighborhood $W \subset U$,

$$T_W M = E_1 \oplus E_2 .$$

We can then (by continuity of the derivative) pick a smaller neighborhood $V \subset W$ on which the tangent map of f is represented by the matrix

$$\begin{pmatrix} A & B \\ C & D \end{pmatrix}$$

satisfying the conditions of lemma 7 for $\varepsilon/2$. Finally, we can clearly pick $\mathcal{N}$ a C^1 (or C^K) neighborhood of f such that the tangent map of g on V differs little enough from $T_V f$ that the conditions hold for ε; and skewness is clearly continuous, so we are done. //

The main result of this section, then, is a strengthening of theorem 1, chapter 5 : besides banning "implosions", it also forbids local "explosions" — in the sense that any new invariant sets will have to range far from Λ.

Theorem 2: If f is as above, and $\varepsilon > 0$, then there exists a compact neighborhood V of Λ in U and a C^1-neighborhood $\mathcal{N}$ of f such that for any pair of diffeomorphisms g_1, $g_2 \in \mathcal{N}$ the following statements hold:

(1) g_i has a unique maximal hyperbolic set Λ_i in V containing every subset of V invariant under g_i.

(2) There is a unique conjugating homeomorphism $h: \Lambda_1 \longrightarrow \Lambda_2$ uniformly within ε of id.

(3) h depends continuously on $(g_1, g_2) \in \mathcal{N} \times \mathcal{N}$

(4) Λ_i has a stable manifold system of size $\geq \beta - \varepsilon$

(5) There exists a continuous family θ of C^K-diffeomorphisms

$\theta_x: W_x(g_1) \longrightarrow W_{h(x)}(g_2)$

between the stable manifold systems.

Proof

The first three statements follow from proposition 4 and a closer examination of the theorem this generalizes (p. 163). Note that it will suffice to prove our present theorem with $g_1 = f$. The theorem alluded to before says that $g_2 = g$ will have an invariant set Λ_g homeomorphic to Λ_1, and clearly $\Lambda_g \subset \Lambda_2$. We can also invoke the same result to say that any map $\tilde{g}$ sufficiently near g will have an invariant set $\Lambda_{\tilde{g}}$ homeomorphic to Λ_2. If we could show that f is close enough to g to qualify as a $\tilde{g}$, we would have by uniqueness that h is onto Λ_2. But the distance allowed depends only on the skewness of g at Λ_2 and on the rate at which T_g

near Λ drops off from this skewness. Invoking proposition 4, we can make the skewness of g on Λ_2 very near the skewness of f on Λ_1, by making g C^1-near f. In fact, by restricting attention to a small neighborhood V of Λ_1, we can insure that Tf on V has skewness very near that on Λ_1 (since f is C^1); then, if Tg is very near Tf, the skewness of Tg will not change much over V, either. So we can assume that any map "g" C^1-within δ_1 of a map "f" with skewness at "Λ_1" σ, near τ (less than ε from τ), has "h": "Λ_1" $\longrightarrow$ "Λ_g". Then by restricting ourselves to a C^1-ball of radius at most $\frac{1}{2}\delta_1$, we see that f will also be within δ_1 of g, and by making g near f, we are sure that g also qualifies as the "f" in the above statement, so there is a conjugacy $h_2: \Lambda_2 \longrightarrow \Lambda_1$ into, and so $h_1(\Lambda_1) = \Lambda_2$.

To prove the statements about stable manifolds, we revise the proof of the stable manifold theorem for sets, this time looking at the stable and unstable manifolds of the operator

$$\phi_{fg} : C(\Lambda_1, U) \longrightarrow C(\Lambda_1, U)$$

$$\phi_{fg}(h) = g \cdot h \cdot f^{-1}.$$

If the stable manifold at h of ϕ_{fg} is

$$\mathcal{W}_g = W^s(\phi_{fg}),$$

then we show as before that these project down to the manifold, this time to give

$$W_{y,g} = \{h(x) \mid h \in \mathcal{W}_g\}$$

and that

$W_{y,g} = W^s_g(y = h(x))$.

The details of mimicry are left to the reader. //

6.4 THE NO-CYCLE PROPERTY AND FILTRATION

Our concepts and arguments have so far been essentially local, concerning the derivative at the non-wandering set, the local structure of Ω , and local stability of basic sets. However, as Palis' example in the last section shows, purely local arguments will not suffice to establish Ω -stability. In this section we deal with the second, global assumption necessary for proving Ω -stability.

Our troubles, in the example, arose roughly from the fact that, for any point on $W^s(C)$, say (see figure 1, p. 207) there were nearby pairs of points such that the ω -limit set of one intersected the α-limit of the other. In other words, for any neighborhood U of such a point, $O_+[U]$ and $O_-[U]$ came arbitrarily close to each other. Thus, it took only a gentle push to "close the gap" and make them intersect, causing Ω -explosions.

This situation arose because we could follow the upper edge of $W^s(C)$ and then (by lemma 1) the lower edge of $W^u(C) = W^s(C')$ under a forward orbit, while under a backward orbit we could follow the <u>lower</u> edge of $W^s(C) = W^u(C')$ to come near $W^s(C')$. Thus, in a very loose sense, there were wandering points (eg, $W^s(C)$) which came very close to going in circles. Now A, B, C, A', B', and C' in the example could have been basic sets just as well as fixedpoints, and then clearly the creation of new basic sets could arise from the analogous situation. This motivated the notion of a "cycle". More precisely,

<u>Definition</u>: If f satisfies Axiom A, then an <u>n-cycle</u> on $\Omega(f)$ is a sequence of basic sets

$$\Omega_o, \Omega_1, \ldots, \Omega_{n+1}$$

with the properties

(1) $\Omega_o = \Omega_{n+1}$

(2) $\Omega_i \neq \Omega_k$ otherwise for $j \neq k$

(3) $W^s(\Omega_i) \cap W^u(\Omega_{i+1}) \neq \emptyset$

Palis shows in [3] that an Ω -stable diffeomorphism satisfying Axiom A cannot have any n-cycles for $n \geq 1$. He shows this by abstracting the example we gave earlier. Thus, under the assumption of Axiom A , a necessary condition for Ω -stability is the

<u>No-Cycle Property</u>: f has no n-cycles with $n \geq 1$.

We shall see that this property (assuming Axiom A) is also sufficient for Ω -stability.

Smale's original formulation of the Ω-stability theorem in [3] had a stronger hypothesis:

<u>Axiom B</u>: If Ω_i and Ω_j are distinct basic sets with

$W^s(\Omega_i) \cap W^u(\Omega_j) \neq \emptyset$, then there exists $p,q \in \text{per}(f)$, $p \in \Omega_i$, $q \in \Omega_j$ with $W^s(p)$ and $W^u(q)$ intersecting transversally at some point.

In the proof sketched in [3], however, one step (proposition 8.4) is the implication Axiom B $\Longrightarrow$ no-cycle property, and from there the proof really only uses no cycles.

An alternate formulation of the no-cycle property is gotten by defining an inequality relation on basic sets by saying

$\Omega_i \leq \Omega_j$ iff

$W^u(\Omega_j) \cap W^s(\Omega_i) \neq \emptyset$

(intuitively, everything flows toward the "lower" elements). The no-cycle property is then just the property that "$\leq$" be a partial ordering of the basic sets.

The last major step in the proof of the Ω-stability theorem, then, is showing more precisely that things really do "flow down" in the ordering. This is done by the filtration, a construction very similar to the handlebody techniques of differential topology. We will, in effect, piece together the manifold M by "climbing up" the ordering: starting with the lowest element, then swallowing up the next element together with its unstable manifold, and so forth, until we reach the top and all of M has been taken in; at each stage, we will be able to assert that no new bits of Ω are introduced within the swallowed part by perturbing f.

The basis of our construction is the following technical

lemma, which speaks only of a basic set and its unstable manifold (so doesn't assume no cycles) but will serve as an induction step in the main theorem.

Lemma 8 (Smale [5], 4.1): If Ω_ℓ is a basic set and U_ℓ a neighborhood of Ω_ℓ, suppose there exists a compact set $X \subset M$ such that

(1) $f[X] \subset \text{int } X$

(2) $O_-[X] \cup \Omega_\ell \supset W^u(\Omega_\ell)$

then there exists an integer $m > 0$ and compact set $Y \subset M$ such that

(1) $f[Y] \subset \text{int } Y$

(2) $\Omega_\ell \subset \text{int } Y \sim f^{-m}[X] \subset U_\ell$

The theorem can be visualized in the following terms: suppose for a moment $\Omega(f)$ is finite and dim $M = 2$, so that Ω_ℓ is a point — assume a saddle, and consider for a moment only the two sinks A, A' that $W^u(\Omega_\ell)$ falls into, and the two sources, B, B', that $W^s(\Omega_\ell)$ originates from. X could then represent a pair of discs surrounding A and A'; we would like to extend X to a new compact set (ultimately, $X \cup Y$) which maps into itself and contains all of $W^u(\Omega_\ell)$. However, we have a further restriction: we want all of that part of Y not accounted for by some finite set of (negative) iterates of X to lie close to Ω_ℓ — ie, in a pre-assigned neighborhood U_ℓ. The situation is sketched in figure 3.

Proof of Lemma 8

We can, without loss of generality, assume U compact and small enough that, using the equivalence of various characterizations

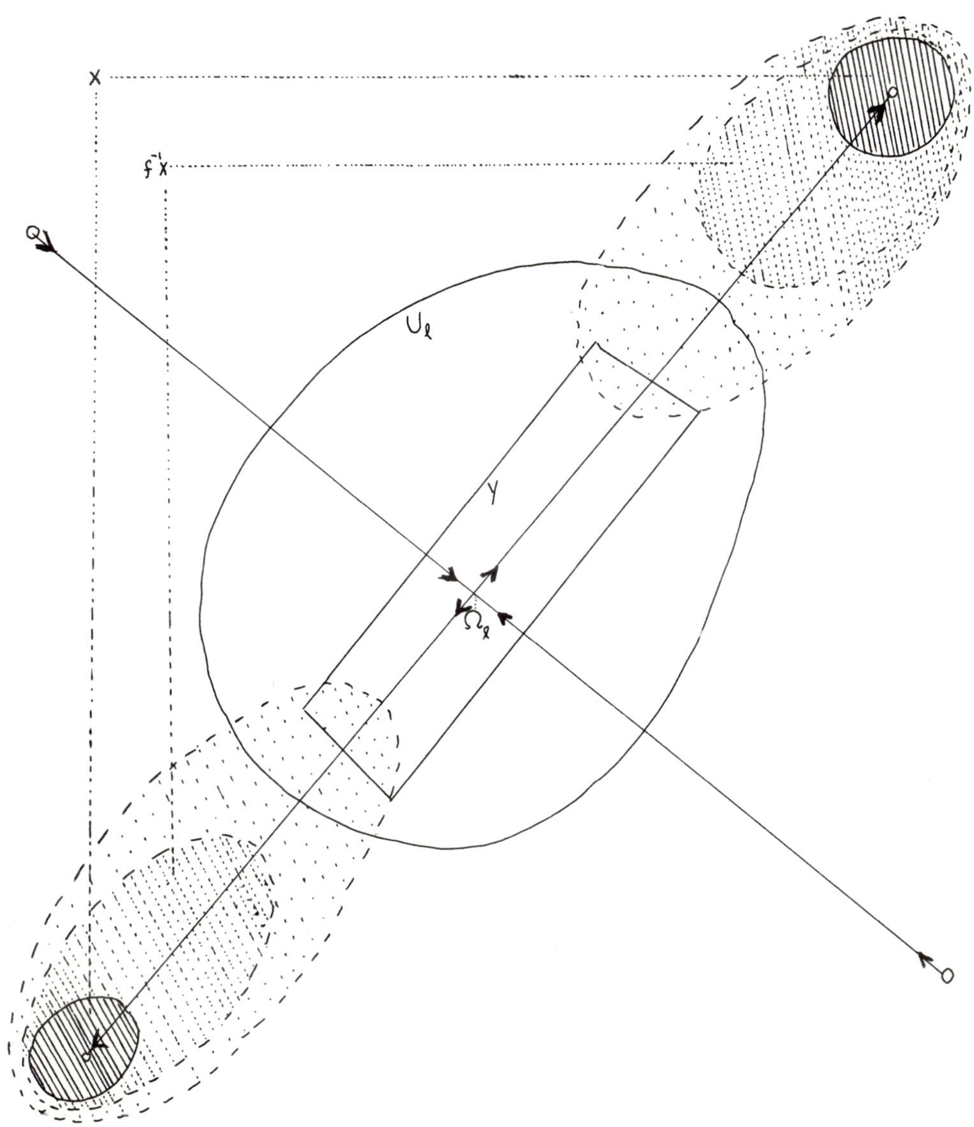

Figure 3: Lemma 8.

of $W^s(\Omega_\ell)$ established in 6.2, we can say

$$\bigcap_{K \geq 0} f^K[U_\ell] \subset W^u(\Omega_\ell).$$

Now, again using 6.2, we can find a compact fundamental domain F^u for Ω_ℓ with:

(1) $\bigcup_{K > 0} f^{-K}[F^u] \subset U_\ell$

(2) $(\exists\, m > 0) \quad f^{-m}[X] \supset F^u$

Define two sets, P and Q, by

$$P = W^u(\Omega_\ell) \cup \bigcap_{K > 0} f^K[X]$$

$$Q = U_\ell \cup f^{-m}X$$

Then we claim: $\bigcap_{K \geq 0} f^K[Q] \supset P$:

Since $f^{-m}[X] \supset \bigcup_{j \leq m} f^{-j}[X]$ (because $fX \subset X$), $f^{-m}X$ contains the positive orbit of F^u; but U_ℓ contains the negative orbit of F^u, so

$$Q = U_\ell \cup f^{-m}[X] \supset O[F^u]$$

and hence each iterate of Q contains $O[F^u]$, so

$$\bigcap_{K > 0} f^K[Q] \supset O[F^u] = W^u(\Omega_\ell) \sim \Omega_\ell.$$

But $Q \supset f^{-m}X \supset X$, so

$$\bigcap_{K > 0} f^K Q \supset \bigcap_{K > 0} f^K X.$$

Finally, we claim: there exists a compact set V satisfying

(1) $Q \supset V \supset \operatorname{int} V \supset P$

(2) $f[V] \supset \operatorname{int} V$.

This is a corollary of the previous claim. Let

$A_r = \bigcap_{r \geq m \geq 0} f^m[Q]$. Then the A_i's form a nested sequence of compact sets converging to P. We claim that for large r, $fA_r \subset Q$: otherwise, $fA_r \sim \text{int } Q$ form a nested sequence of compact sets, and their intersection has the property that its inverse image lies in $\bigcap A_r = P$, but it is disjoint from P, contradicting invariance of P. But if $fA_r \subset Q$, then clearly

$fA_r \subset A_{r+1}$

and we can assert that there is an $n \geq 2$ so that

$f^n A_r \subset \text{int } A_r$.

We would like to reduce n to 1. We do this by "downward induction" on n , as follows: Pick compact W such that

(1) $A_r \subset \text{int } W$

(2) $f^n W \subset \text{int } A_r$

Then define

$E = A_r \cup (W \cup f^{n-1}W)$.

We need to show $f^{n-1}[E] \subset \text{int } E$. Well,

$f^{n-1}A_r \subset \text{int } f^{n-1}W$

$f^{n-1}A_r \subset A_{r+1} \subset A_r \subset \text{int } W$

so $f^{n-1}[A_r] \subset (\text{int } W) \cap (\text{int } f^{n-1}W) \subset \text{int } E$. Also, (remember, $n \geq 2$)

$$\begin{aligned} f^{n-1}[W \cap f^{n-1}W] &\subset f^{2n-2}W \\ &= f^{n-2} f^n W \\ &\subset f^{n-2} \text{ int } A_r \\ &\subset \text{int } A_r \subset \text{int } E \end{aligned}$$

so

$f^{n-1}E \subset \text{int } E$.

We have shown the existence of a compact neighborhood in Q of P with $fV \subset \text{int } V$, so that $\bigcap_{m \geq 0} f^m V = P$; now let $Y = V \cup f^{-m}X$, and clearly $fY \subset \text{int } Y$. Also, since $Y \subset Q$, $\text{int } Y \sim f^{-m}X \subset U_\ell$. To see that $\Omega_\ell \subset \text{int } Y \sim f^{-m}X$, it suffices to see that $\Omega_\ell \subset \text{int } V$, and this is because $\Omega_\ell \subset P$. //

We can now proceed to define and construct the filtration. As before, we order the basic sets by

$\Omega_i < \Omega_j$ if $W^s(\Omega_j) \cap W^u(\Omega_i) = \emptyset$

(ie, $\Omega_i < \Omega_j$ if $\Omega_j \not\leq \Omega_i$). We can, if necessary, extend this to a linear ordering, and then renumber so that

$\Omega_o < \Omega_1 < \cdots < \Omega_K$.

We pick disjoint neighborhoods $U_i \supset \Omega_i$ by 6.2, satisfying

$$\bigcap_{m \in \mathbb{Z}} f^m[U_i] = \Omega_i$$

$$\bigcap_{m > 0} f^m[U_i] = W^u(\Omega_i)$$

$$\bigcap_{m > 0} f^{-m}[U_i] = W^s(\Omega_i) \quad .$$

The filtration is then given by

Theorem 3 (Filtration): If f satisfies Axiom A and the no-cycle property, then there exists a nested sequence of compact sets

$$M_o \subset M_1 \subset \cdots \subset M_K = M$$

and a sequence of positive integers

$m_1, \ldots, m_K$

such that

(1) $f^{-m_\ell}(M_{\ell-1}) \subset \text{int } M_\ell$

(2) $f(M_\ell) \subset \text{int } M_\ell$

(3) $\Omega_\ell \subset \text{int } M_\ell \sim f^{-m_\ell} M_{\ell-1} \subset U_\ell$

(4) $M_\ell \subset \bigcup_{i \leq \ell} W^s(\Omega_i)$.

<u>Proof</u>

The proof is by induction on ℓ : (1) $\ell = 0$:

We note that $W^u(\Omega_0) \cap W^s(\Omega_i) = \emptyset$ for any $i > 0$, so that (since $M = \bigcup W^s \Omega_i$) $W^u \Omega_0 = \Omega_0$. (The lowest element in the order is an "attractor".) In lemma 8, we let $X = \emptyset$ to get $M_0 = Y$, possessing all of the required properties by lemma 8.

(2) The induction step:

Given $M_{\ell-1}$, we need only check that

$$\bigcup_{r>0} f^{-r}(M_{\ell-1}) \cup \Omega_\ell \supset W^u(\Omega_\ell)$$

to apply lemma 8, with $X = M_{\ell-1}$.

The inclusion follows from two others: First,

$$W^u(\Omega_\ell) = \bigcup_{i=0}^{K} [W^u(\Omega_\ell) \cap W^s(\Omega_i)]$$

$$= \bigcup_{i \leq \ell} [W^u(\Omega_\ell) \cap W^s(\Omega_i)]$$

the former equality follows from the corollary of the spectral decomposition theorem (p. 191), the latter from the no-cycle property. Also,

$$\bigcup_{m>0} f^{-m}(M_{\ell-1}) \supset W^s(\Omega_i) \qquad (\forall\, i > \ell)$$

by construction. So we apply lemma 8 to get M_ℓ satisfying (1)-(3), and (4) follows by induction on ℓ and the choice of U_i, giving us our filtration. //

6.5 GLOBAL STABILITY

After all of this work, the Ω-stability theorem becomes a fairly easy result.

Theorem 4 (The Ω-Stability Theorem): If $f \in \mathrm{Diff}(M)$ satisfies axiom A and the no-cycle property, then f is Ω - stable.

Proof

We construct a filtration $\{M_\ell\}$ as in 6.4, being careful to pick the U_ℓ's so as to satisfy local stability (theorem 2, 6.3). Then:

(1) Claim: There exists a C^0 neighborhood $\mathcal{N}_0$ of f such that any $g \in \mathcal{N}_0$ satisfies

$$\Omega(g) \subset \bigcup_{\ell=0}^{K} (\operatorname{int} M_\ell) \sim f^{-m_\ell}(M_{\ell-1})$$

$$\subset U_\ell .$$

This is because for g near f, $g^{m_\ell}(f^{-m_\ell} M_{\ell-1}) \subset M_{\ell-1}$.

(2) We write $\Omega(g)$ as a union

$$\Omega(g) = \bigcup_{\ell=0}^{K} \Lambda_\ell$$

where $\Lambda_\ell = \Omega(g) \cap [\operatorname{int} M_\ell \sim f^{-m_\ell} M_{\ell-1}]$. Since $\Lambda_\ell \subset U_\ell$, by theorem 2 all g in a C^1-neighborhood $\mathcal{N}_1$ of f will have

$h_\ell : \Omega_\ell \longrightarrow \Lambda_\ell$

conjugating $f|\Omega_\ell \sim g|\Lambda_\ell$.

(3) Obviously, the h_ℓ's together define a global

$h: \Omega(f) \longrightarrow \Omega(g)$

and so

$g \underset{\Omega}{\sim} f$

for $g \in \mathcal{N}_0 \cap \mathcal{N}_1$. //

CHAPTER 7

A SURVEY OF RECENT WORK

The main concern of these notes has been to give in some detail the ideas leading up to Smale's Ω-stability theorem, and thus our development has more or less followed the outlines of Smale's survey [3], published in 1967. This chapter will serve to briefly sketch a few areas of interest that have developed since then, including the resolution of several questions posed in [3]. The reader is warned that the judgement used in selecting topics here is highly subjective, and the treatment very superficial.

We shall group most of our survey around two broad topics: global stability questions, and local analysis of invariant sets.

7.1 STABILITY QUESTIONS

The reader will recall from chapter 0 that a major motivation for introducing seemingly artificial notions, such as hyperbolicity, was the hope that eventually a "generic" classification scheme might arise to describe "most" diffeomorphisms. This hope was of course strengthened by the early discovery of generic properties; the Kupka-Smale theorem,

Peixoto's work on the circle. The Ω-stability theorem, too, was originally presented as a candidate for a "generic" result.

A series of counterexamples in the last few years has, however, demonstrated that we are farther from a completion of the "generic" program than seemed the case in the innocence of 1965.

Peixoto's results seemed to indicate a first candidate for genericity - the Morse-Smale systems, which he showed open and dense on the circle. Thom's discovery of hyperbolic toral automorphisms, however, pointed out that generic descriptions in higher dimensions couldn't be that simple. The possibility remained, though, that a more abstract generalization of Peixoto's results might hold - perhaps structural stability might be generic.

This conjecture was shot down for dimensions greater than 3 by Smale in [6]. He exhibited an open set of diffeomorphisms of the 3-torus (or of any 3-manifold) such that arbitrarily near each element were others not conjugate to it.

His example rests on the "DA" ("derived from Anosov") diffeomorphisms of the 2-torus. These are derived from a hyperbolic toral automorphism by modifying the map on a neighborhood of a fixedpoint, replacing it with a pair of saddles and a sink:

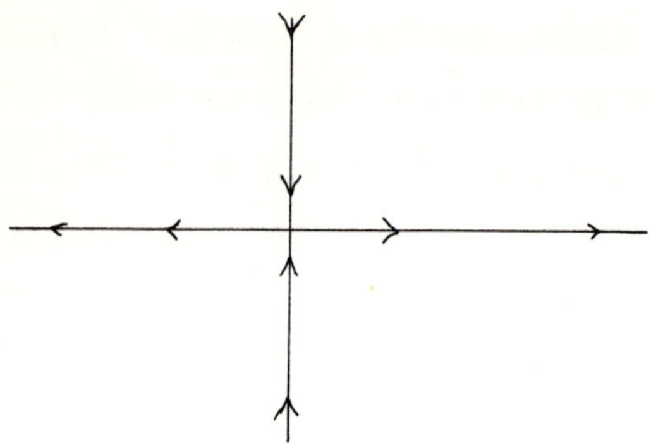

Figure 1a: original fixedpoint

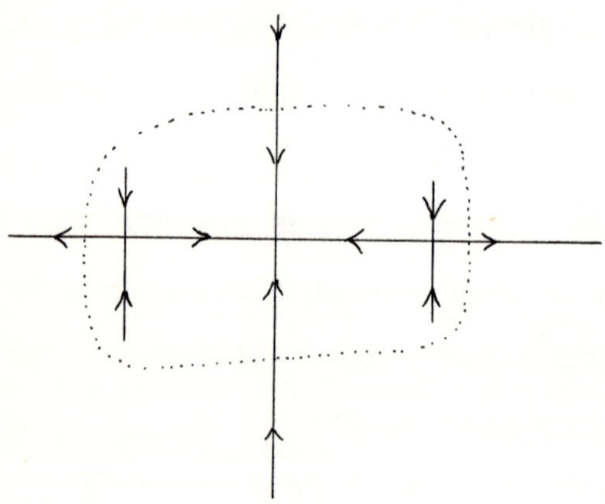

Figure 1b: modified map

The resulting map on the torus will have the stable manifolds of the saddles bounding the stable manifold of the sink, which winds around T^2 asymptotic to the original stable manifold of the original fixedpoint:

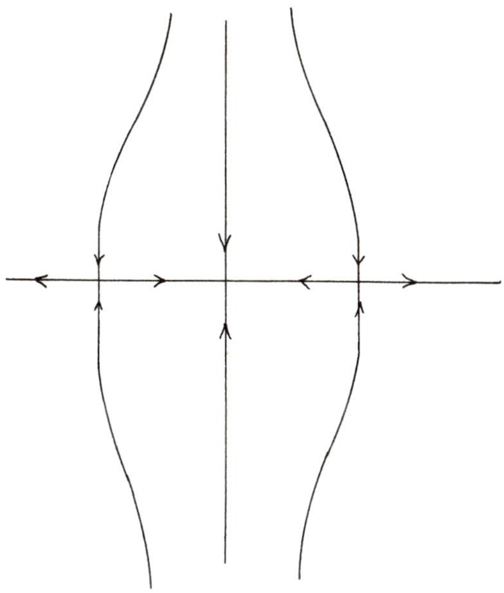

Figure 1c: Stable manifolds in the DA .

The non-wandering set of the DA will consist of the sink together with the closure of the stable manifolds of the saddles, which will form a "solenoid".

Smale's example is then gotten by crossing T^2 with an interval, and making the map on $T^2 \times I$ an isotopy between the original hyperbolic map and the new DA, modified so that the only invariant fibers of the projection on I are those at the two endpoints. The instability will arise from the fact that stable and unstable manifolds may intersect in different ways.

The example was slightly modified by R. F. Williams [1] to produce a similar example on the 2-torus itself. It was also extended by Peixoto and Pugh [1] to non-compact manifolds.

These examples, however, motivated the formulation of the Ω -stability theorem; in Smale's survey [3] axiom A and hence Ω -stability were offered as new candidates for generic properties.

In 1968, however, examples due to Abraham and Smale [1] and Shub (Hirsch, Pugh, Shub [1]) showed that in dimensions four and greater, neither axiom A nor Ω-stability are generic. Again, the examples involved an open set of diffeomorphisms such that near each element were others not Ω-conjugate to it. The Abraham-Smale example uses an idea similar to the Smale structural stability example: a map of $T^2 \times S^2$ to itself with the tori over different fixedpoints of S^1 corresponding to different Anosov or DA maps.

In the summer of 1968, Newhouse [2] introduced a different idea to construct an example of Ω -instability on surfaces, and hence on all non-1-dimensional manifolds. His construction is a modification of the "horseshoe" discussed in chapter 4. Quite simply, he adds a "hook" to the image of the rectangle, so that in addition to passing through R twice with a "U-turn" outside, fR passes through R a third time, this time insolently making its U-turn inside R:

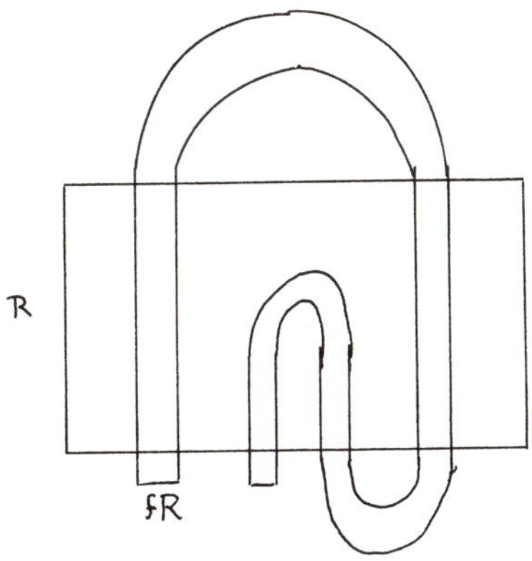

Figure 2: Newhouse's Example.

This turn inside R creates a tangency of stable and unstable manifolds, which destroys hyperbolicity, and creates Ω -havoc by alternately occuring on stable manifolds of periodic or of non-periodic points. The technical verification of this behavior, of course, is not as simple as it sounds. Recently, C. Simon [1] applied the Smale isotopy idea to the Newhouse hook to create an example on T^3 which is rich in its implications.

In the light of these examples, there has been a lessening of the search for a generic notion of stability, and more attention seems to have been focused on deeper analysis of already existing notions.

The analysis of structural and Ω-stability seems now to take the Ω-stability theorem as a first step in a search for necessary and sufficient conditions for stability. The conditions conjectured (originally by Smale) for Ω-stability are naturally motivated by the Ω-stability theorem:

Conjecture 1: Ω-stability $\Longleftrightarrow$ Axiom A + no-cycle property.

The Ω-stability theorem establishes one direction, and Palis' work on explosions reduces the other direction to necessity of Axiom A for Ω-stability. As mentioned before, J. Franks has made some progress in this direction, showing ([3]) that all periodic points must be hyperbolic, and ([4]) that a strengthened version of Ω-stability is equivalent to Axiom A plus no-cycles.

To formulate the corresponding conjecture on structural stability, we need a

Definition: A diffeomorphism satisfying axiom A has the strong transversality property if every stable manifold meets every unstable manifold transversally.

The structural stability conjecture, then is,

Conjecture 2: Structural Stability $\Longleftrightarrow$ Axiom A + strong transversality.

Recently, J. Robbin [1] has proven half of this.

Theorem: In $\mathrm{Diff}^2(M)$, Axiom A + strong transversality $\Longrightarrow$ structural stability.

His proof proceeds in three large steps. A map is infinitesimally stable if the operator $(1 - f^{\#})$, where $f^{\#}$ is the usual action of f on vectorfields, has a right inverse on an appropriate space of vectorfields. Robbin first shows that infinitesimal stability implies structural stability. (There is some belief that the converse may also hold.) Then he defines a notion of "locally Anosov" and shows that this property on one hand implies infinitesimal stability and is on the other hand implied by C^2, axiom A, and strong transversality.

A proof of conjecture 1 would presumably imply conjecture 2, since structural stability implies Ω-stability, and strong transversality seems, on first glance at least, necessary for structural stability.

Several other notions of stability have been proposed (future stability, tolerant stability) but these don't seem to have proven very fruitful. Shub and Williams [1], for example, show that future stability is not generic.

An interesting sidelight to these stability questions is a recent paper of Franks [5], investigating conditions for a diffeomorphism to be C^K-stable - that is, for perturbations to be conjugate via smooth homeomorphisms.

Finally, an important question related to these concerns is the logical independence of the two parts of Axiom A. Smale in [3] asks

Question: Does A(a) imply A(b) - ie, does hyperbolic structure on $\Omega(f)$ imply density of periodic points in $\Omega(f)$?

A theorem in the folklore, attributed to Anosov, states that the periodic points are dense among the non-wandering points for $f|\Omega(f)$. More recently, Newhouse [3] has made some significant progress toward an affirmative answer to this question.

7.2 ANALYSIS OF INVARIANT SETS

Another direction taken by the new emphasis on analysis of existing notions has been a closer study of the structure of Ω under the assumption of axiom A. Such studies could be grouped under three headings: (1) a further development of the perturbation theory of invariant sets, (2) an attempt to study the topological structure of basic sets, and (3) a more precise analysis of the dynamic structure of basic sets.

The first heading could be regarded as starting from the stability theorems for hyperbolic fixedpoints (theorem 3, chapter 2) and for hyperbolic sets (theorem 1, chapter 5, theorem 2, chapter 6). The subsequent work has dealt primarily with invariant manifolds.

The first result in this direction was proven by Sacker [2].

Definition: Let $f \in \mathrm{Diff}(M)$, and suppose $V \subset M$ is a compact submanifold of M invariant under f: $f[V] \subset V$. We say f is normally r-hyperbolic at V if the tangent bundle of M at V splits into an invariant Whitney sum

$$T_V M = TV \oplus N^u \oplus N^s$$

where TV is the tangent bundle to V, and Tf satisfies:

(1) Tf contracts N^s more than TV:

$$\sup_{x \in V} \| Tf | N^s_x \| < \inf_{x \in V} m(Tf^r | T_x V)$$

(2) Tf expands N^u more than TV:

$$\inf_{x \in V} m(Tf|N_x^u) > \sup_{x \in V} \| Tf^r | T_x V \| .$$

Recall that $m(L)$ stands for the minimum norm of L: $\| L^{-1} \|^{-1}$.

Sacker's theorem can then be stated as:

Theorem: If f is normally r-hyperbolic at V, then
(1) V is a C^r-submanifold of M
(2) Any C^1-perturbation g of f will have a normally r-hyperbolic submanifold V_g near V.

A refinement of this result is given by Fenichel [1].

In a forthcoming paper, Hirsch, Pugh and Shub [1] give a new proof of Sacker's theorem, proving analogues of Hartman's theorem and the stable manifold theorem for normally r-hyperbolic submanifolds. They than can generalize the result to Banach manifolds, using it to prove

Theorem (Equivarnant Fibration Theorem): If $\pi: E \longrightarrow M$ is a fibration such that the fibers form a continuous family of C^r manifolds and if $F: E \longrightarrow E$ is a fiber map covering $f: M \longrightarrow M$ with:
(1) f Anosov of skewness
(2) there is a family of normal bundles to the fibers, invariant under TF and hyperbolic with skewness $> (\tau)^{\frac{1}{r}}$ then any C^1 perturbation G of F will have a nearby fibration $\pi_G: E \longrightarrow M$ so that G is a fiber map covering f.

The second area of concern, the topological structure of basic sets, was posed as a question in Smale's survey [3]. He conjectured at that time that locally, every basic set is the product of a submanifold with a cantor set. In his thesis,

J. Guckenheimer [1] presents a counterexample to this conjecture. His example results from a study of maps $S^2 \longrightarrow S^2$ (regarded as the compactified complex plane), given by the quotient of two polynomials. He employs a construction, due to R. F. Williams, which allows one to realize the inverse limit of an expansive map as the restriction of a diffeomorphism to an invariant set. Briefly, this technique consists of taking $f: A \longrightarrow A$, using the Whitney embedding theorem to approximate f by an embedding into a high-dimensional manifold, and extending to a map F on a tubular neighborhood N which strongly contracts fibers. Then the inverse limit, $\varprojlim f_a : K \longrightarrow K$, abstractly defined as a subshift in the direct product of a bisequence of copies of A by

$K = \{\{x_i\}_{i \in \mathbb{Z}} \mid x_i \in A, \ fx_i = x_{i+1}\}$

can be realized as

$F|(\bigcap_{n \geq 0} F^n[N])$.

Williams ([2], [3], [4], [5]) has used this construction and a generalization of Shub's expanding maps to develop a classification of "attractors" - those basic sets which appear at the bottom of the ordering, characterized by

$W^u(\Lambda) = \Lambda$.

The structure is a generalization of the "solenoids" of classical mechanics.

Finally, Hirsch [2] has some other topological structure

theorems for invariant hyperbolic sets.

The third heading, the dynamical structure of basic sets, has concerned itself with two kinds of questions.

The first is the rationality of the zeta function. Smale [3] proposed the following device for counting the periodic orbits of a diffeomorphism, following Artin and Mazur [1].

Definition: Let f be a Kupka-Smale diffeomorphism of compact manifold M, and denote by N_m the number of fixedpoints of f^m. Then define the zeta function of f as the complex power series

$$\zeta(t) = \exp \sum_{m=1}^{\infty} \frac{N_m}{m} t^m .$$

Artin-Mazur [1] showed that this series has a positive radius of convergence for a dense set of Diff (M), M compact. Smale then suggested that $\zeta(t)$ might be rational (quotient of polynomials) for generic f. This would mean $\zeta(t)$, and hence the various numbers of periodic points, would be determined by a finite number of invariants. Guckenheimer [2] shows that Axiom A together with the no-cycle property implies $\zeta(t)$ rational; more recently, Manning [1] has shown Axiom A alone implies $\zeta(t)$ rational. On the other hand, Simon [1] has shown that any residual set of diffeomorphisms will include some with non-rational $\zeta(t)$.

Other results related to the counting of periodic points have been obtained by Banchoff and Rosen [1], and Sondow [1] for Anosov maps, and by K. Meyer ([1], [2]) and Shub [2] for

Axiom A systems.

A second important development in the dynamics of basic sets has been Bowen's investigation ([1]-[4]) of the relation between basic sets and shift automorphisms (see section 4.1), with implications for questions of both topological dynamics and ergodic theory on such sets. In [1], he calculated the topological entropy of a diffeomorphism satisfying axiom A by counting periodic points; he also showed there that any zero-dimensional basic set is conjugate to the restriction of the shift to an invariant set defined by excluding certain "blocks" from admissible members. These are the "subshifts of finite type" often considered in topological dynamics. A result of Parry and of Adler and Weiss says that these subshifts have a certain invariant measure with nice properties. In [2], Bowen again used a counting procedure on periodic points to construct an invariant measure on any basic set Λ which:

(1) vanishes on points

(2) is positive on open sets of Λ

(3) is ergodic for f.

In [3], he constructs a "Markov partition" for any basic set. This is a finite cover $C = \{A_1, \ldots, A_r\}$ of Λ by open subsets satisfying:

(1) each A_i is covered by the "canonical coordinates" of

6.1

(2) each A_i is the closure of its interior in Λ.

(3) if $i \neq j$, $A_i \cap A_j \subset \partial A_i \cap \partial A_j$

(4) whenever $x \in \mathrm{int}_\Lambda A_i \cap f^{-1}\mathrm{int}_\Lambda A_j$,

$$\begin{cases} f(A_i \cap W^u_{loc}(x)) \supset A_j \cap W^u_{loc}(f(x)) \\ f(A_i \cap W^s_{loc}(x)) \subset A_i \cap W^s_{loc}(f(x)). \end{cases}$$

The notion was first introduced by Sinai [1].

A Markov partition then allows Bowen to introduce symbolism and construct a map $\pi:\Sigma \longrightarrow \Lambda$, where Σ is again a subshift of finite type. This map is:

(1) a surjective semiconjugacy.

(2) a measure-theoretic isomorphism between the shift and $f|\Lambda$.

In the measure-theoretic case, the measure on Σ is a natural one given by a theorem of Parry, and the measure on Λ is given by Bowen's generalization of Parry's result:

<u>Theorem</u>: Λ has a unique normalized f-invariant Borel measure μ_f whose entropy equals the topological entropy of f.

Most recently, in [4], Bowen continues to study the map π, proving that the cardinality of "fibers" is bounded and that it sets up correspondences for periodic points, minimal sets, and transitive orbits between Σ and Λ. As a consequence, he can conclude

<u>Theorem</u>: Minimal sets of Λ have dimension 0.

Earlier constructions of invariant measures were given by Sinai [1] for Anosov maps, and Krzyżewski-Szlenk [1] and

Hirsch [1] for expanding maps. A related analytic tool is Williams' characterization of "shift-equivalence" for subshifts of finite type in [6]. A more detailed survey of some of this work is given by Baxendale [1].

7.3 OTHER AREAS

Several interesting lines of enquiry have developed from the application of the ideas discussed in these notes to various phenomena in differential equations.

The theory of Hamiltonian systems, of course, arises from classical mechanics and is an old field of research. There has recently been the development of a "generic" theory of Hamiltonian fields. This was initiated by Abraham and Marsden in [1], who outlined a possible program for such a study. R. C. Robinson ([1], [2]) answered many of the questions posed by Abraham and Marsden, and has developed the beginnings of this generic theory, particularly formulating and proving appropriate modifications of the Kupka-Smale theorem for Hamiltonian, symplectic, and volume-preserving diffeomorphisms and flows. Pugh and Robinson [1] have recently established the C^1-closing lemma for various classes of diffeomorphisms, including the above types.

A study of generic properties for second-order, non-conservative (dissipative) flows has been made by

Shahshahani [1].

A second interesting area has been the study of Anosov flows, which seems to have developed along lines slightly different from the corresponding diffeomorphisms. Recent work has included the theses of Byers [1], Tomter [1], and Plante [1], [2].

The recent generalization of the notion of Anosov flows to actions of $\mathbb{R}^n$ by Hirsch, Pugh, and Shub has revived interest in actions of more general groups than $\mathbb{Z}$ or $\mathbb{R}$. Discussion of all the aspects of earlier work on group actions on a compact manifold would take us too far afield here ; the reader is referred to Smale's survey [3] for references. We mention , however, a few papers which fit into the general approach we have been considering. Kopell [1] in her thesis studied various problems concerning pairs of commuting diffeomorphisms, giving in particular some "generic" properties of fixedpoints for such actions of the circle. Schneider [1] in his thesis established a version of the Kupka-Smale theorem for actions of $SL(n, \mathbb{R})$ on a compact manifold. Fields [1] has proven a Kupka-Smale theorem for vectorfields which commute with a compact-group action. Lima, Sacksteder and H. Rosenberg have studied actions of $\mathbb{R}^K$ (ie, sets of commuting vectorfields) on compact manifolds.

The recent work of Hirsch, Pugh, and Shub concerns a special class of $\mathbb{R}^K$-actions, the "Anosov" actions. Roughly, an

$\mathbb{R}^K$-action is Anosov if the diffeomorphism given by some element of $\mathbb{R}^K$ has an invariant splitting $T = E^s \oplus E^u \oplus W$, where W is tangent to the orbits of the whole group, E^s contracts and E^u expands. Such a member of $\mathbb{R}^K$ is called an "Anosov element". Hirsch-Pugh-Shub [1] show such an action, if locally free, is structurally stable. They also prove structural stability in case the group is connected and has Anosov elements in its center, even if the action isn't locally free. Pugh-Shub [1] show that if there exist Anosov elements in the center and some measure is preserved, the action is ergodic. Hirsch [3] obtains an exponential estimate on the growth of an orbit.

Finally, as a corollary of the structural stability theorem, Hirsch-Pugh-Shub show that a vectorfield C^1-near zero which commutes with an Anosov flow Y must be a function-multiple of Y:

$X_m = f_{(m)} \cdot Y_m$.

An interesting question of a very different flavor is posed by Smale's

<u>Conjecture:</u> If M is compact, the group $\text{Diff}(M)$ is a simple group.

The analogous result for the group of homeomorphisms and of PL homeomorphisms of S^1 is known. At the 1969 Summer Symposium on Dynamical Systems at the University of Warwick (Coventry, England), D. Epstein presented the following partial results in this direction (DM = Diff (M)) :

Theorem: (1) The commutator group [DM,DM] is simple
(2) [DM, DM] is dense in DM
(3) Every normal subgroup of DM contains [DM, DM] .

The only written reference I am presently aware of for these results is a summary in the Report of Proceedings of the Symposium.

APPENDIX I

ENDOMORPHISMS OF THE CIRCLE

This appendix expands on section 1.3 ; we shall sketch two alternate proofs of Shub's stability theorem and try to indicate what is known concerning non-expanding maps on S^1. A map $f \in C^r(S^1, S^1)$, when regarded as a dynamical system (as in 1.3), is often referred to as an endomorphism of S^1 ; note that such a map need not be an endomorphism in the group-theoretic sense.

Our arguments in this section will be specific to the circle, and will refer to concepts treated in other chapters of the text.

Recall from 1.3 :

Theorem (Shub): If f and g are expanding maps of the same degree, they are topologically conjugate.

Recall also that we had a set of canonical "models" of expanding maps; the model in degree n was given by the "z^n" map ζ_n, defined by

$$\zeta_n[\exp(t)] = [\exp(t)]^n = \exp(nt)$$

where exp is the projection $\exp(t) = e^{2\pi i t}$. We shall exploit these models, giving two proofs of a fact equivalent to Shub's, namely:

Every expanding map f on S^1 of degree n is conjugate to ζ_n.

<u>Proof I</u>

This proof will be direct and geometric; we will construct an injective semi-conjugacy φ from ζ_n to f.

Begin by noting that ζ_n has a "natural" lifting to $\mathbb{R}$, given by

$$\tilde{\zeta}_n(t) = nt .$$

This lift is a diffeomorphism of $\mathbb{R}$, its inverse is a contraction, and hence it has a unique fixedpoint, 0, which projects to the fixedpoint $1 = \exp(0) \in S^1$ for ζ_n (Caution: ζ_n itself can have other fixedpoints, but these are not reflected in $\tilde{\zeta}_n$). Similarly, f has a lift, F, which has a unique fixedpoint; we lose no generality by assuming it to be 0, so that $f(1) = 1$.

Since a semi-conjugacy takes fixedpoints to fixedpoints, it will be natural to define

$$\varphi(1) = 1.$$

We now consider the inverse image of 1 under each map. The points mapping to 1 under ζ_n are exactly the nth roots of unity - which can be represented in $\mathbb{R}$ by the points

$$0, \frac{1}{n}, \frac{2}{n}, \ldots., \frac{n-1}{n} .$$

Similarly, f, being of degree n, maps exactly n points p_i $(i = 0\ldots n-1)$ to 1; we can number these so that they can be represented on $\mathbb{R}$ by the points in the unit interval

$0 = P_o < P_1 < P_2 < \cdots < P_{n-1}$

where $p_i = \exp(P_i)$. We now define

$$[\exp(\tfrac{m}{n})] = p_m \qquad (m = 0, \ldots, n-1),$$

noting that for $m = 0$, this definition agrees with the first.

Suppose now, inducting on K that we have extended φ to an order-preserving conjugacy from the (n^K)th roots of unity to the set of points $q_o \cdots q_{n^K-1}$ mapping to 1 under f^K.

The Kth inverse image of 1 under ζ_n is

$$\zeta_n^{-K}[1] = \exp\{0, \frac{1}{n^K}, \frac{2}{n^K}, \ldots, \frac{n}{n^K} = \frac{1}{n^{K-1}}, \ldots, \frac{n^K-1}{n^K}\}$$

while the $(k+1)$th inverse image is

$$\zeta_n^{-(K+1)}[1] = \exp\{0, \frac{1}{n^{K+1}}, \ldots, \frac{n}{n^{K+1}} = \frac{1}{n^K}, \ldots, \frac{n^{K+1}-1}{n^{K+1}}\}$$

and we note that the (n^{K+1})th roots of 1 which are not also (n^K)th roots are exactly the exponentials of those $\frac{m}{n^{K+1}}$ for which m is not divisible by n. We can express this in terms of the map ζ_n, as follows: the interval I between $\exp(\frac{m}{n^K})$ and $\exp(\frac{m+1}{n^K})$ maps under ζ_n bijectively to the interval J between $\exp(\frac{m}{n^{K-1}} = \frac{mn}{n^K})$ and $\exp(\frac{m+1}{n^{K-1}} = \frac{mn+n}{n^K})$; J contains exactly $n-1$ other (n^K)th roots of unity:

$$\exp(\frac{mn+1}{n^K}),\ \exp(\frac{mn+2}{n^K}),\ \ldots\ \exp(\frac{mn+n-1}{n^K})$$

and their inverse images in I are exactly the points in I which map onto 1 by ζ_n^{K+1} but not by ζ_n^K.

A similar analysis indicates that between two successive

elements q_m, q_{m+1} of $f^{-K}[1]$ lie $(n-1)$ other elements of $f^{-(K+1)}[1]$, given as the unique inverse images in the interval (q_m, q_{m+1}) of the points of $f^{-K}[1]$ which lie strictly between $f(q_m)$ and $f(q_{m+1})$. We then extend φ to (n^{K+1})th roots of unity by

$$\varphi\left(\frac{mn+j}{n^{K+1}}\right) = \text{the (unique) point } p \in (q_m, q_{m+1}) \text{ such that } f(p) = \varphi\left(\frac{mn+j}{n^K}\right)$$

This is, of course, the definition uniquely determined by the twin conditions of semi-conjugacy and order-preservation. The induction leads to a definition of φ as an order-preserving conjugacy from the set of (iterated) inverse images of 1 under ζ_n (the (n^K)th roots of unity, K ranging over positive integers) onto the set of inverse images of 1 by f. By an easy application of Lemma 1 in 1.3, each of these inverse orbits is dense in S^1, so that φ extends uniquely to a conjugacy on all of S^1. //

Proof II

The second (or third, if we count the one in 1.3) proof is similar to Proof I, except that we employ symbolic dynamics and Markov partitions. We mention in passing that in addition to the work of Sinai and of Bowen on general Markov partitions, Hirsch [1] and Baxendale [1] have done related work on expanding maps, and Krzyżewski and Szlenk [1] have constructed invariant measures for expanding maps.

We start with an analysis of ζ_n and its lift, $\tilde{\zeta}_n$. If we express numbers in powers of n instead of 10, we have corresponding to each real number a sequence

$\ldots a_{-1}\, a_0 \cdot a_1\, a_2 \ldots$

of integers, $0 \leq a_i < n$, with of course the usual ambiguity about sequences with a tail of (n-1)'s. If however we adopt a convention - say that in such cases we pick the terminating sequence ($\{\ldots K\ 0\ 0 \ldots\}$ instead of $\{\ldots (K-1)\ (n-1)\ (n-1) \ldots\}$) then the action of $\tilde{\zeta}_n$ can be expressed as

$\tilde{\zeta}_n(\{a_K\}) = \{a_{K+1}\}$

which is exactly the shift automorphism on n symbols (see 4.1). The projection exp identifies two sequences that agree to the right of the decimal point; thus if we regard S^1 as the reals modulo one instead of as the unit circle in the complex plane, ζ_n becomes just multiplication by n, mod 1, and can be expressed as a "semi-shift" - the analogue of the shift automorphism, but for one-sided sequences of the symbols 0, ..., (n-1).

The analysis of ζ_n so far can be expressed by saying

There is a semi-conjugacy from the semi-shift on n symbols to the action of ζ_n on S^1, which fails to be injective precisely on the sequences that eventually are all zero or all (n-1); the semi-conjugacy identifies one eventually-null sequence with each eventually -(n-1) sequence.

We can now analyze precisely what the semi-conjugacy means in dynamic terms, as follows. Note that a sequence has a "0" in the first place to the right of the decimal if (and only if) its projection lies in the interval between 1 and $\sqrt[n]{1}$ given by $\exp[0, \frac{1}{n}]$. It has a "1" iff it projects to a point in $[\exp(\frac{1}{n}), \exp(\frac{2}{n})]$; it has an "m" in that position iff it projects into the interval $[\exp(\frac{m}{n}), \exp(\frac{m+1}{n})]$. Thus, specification of the first element of the sequence partitions S^1 into n abutting intervals $A_0, A_1, \ldots, A_{n-1}$.

From the semi-conjugacy, we see easily that the second entry specifies, in similar fashion, the location of the image of the projection:

$$\zeta_n[\pi\{a_K\}] \in A_{a_2},$$

We continue this reasoning to see that the ℓ th entry of a sequence specifies the element of the partition containing the $(\ell-1)$th image of the projection. Thus, the projection of a sequence $\{a_K\}$ is defined as the point p for which

$$\zeta_n^{\ell}(p) \in A_{a_{\ell-1}}.$$

This point is uniquely defined because all intervals expand uniformly under ζ_n. In formal terms,

$$\pi(\{a_K\}) = \bigcap_{K=1}^{\infty} \zeta_n^{-(K-1)} A_{a_K}$$

We note that the sequences for which this description is ambiguous are precisely those which are eventually either (n-1) or 0; these project (respectively) to those points that eventually stay in A_0 or in A_{n-1}. But the only point that stays in A_0 under all iterates of ζ_n is 1, and likewise only 1 stays in A_{n-1} — so that the points of ambiguity project exactly to the inverse orbit of 1, and the ambiguity is reflected in the fact that 1 is the common endpoint of A_0 and A_{n-1}.

We now turn to f, and note that the first inverse image of 1 yields a partition $\{B_0, B_1, \ldots, B_{n-1}\}$ of S^1, where $B_i = [p_i, p_{i+1}]$. We note that by reversing the thought process above we can define a semi-conjugacy from the semi-shift on $(n-1)$ symbols to f by

$$\pi_f(\{a_K\}) = \bigcap_{K=1}^{\infty} f^{-(K-1)} A_{a_K} .$$

We leave to the reader the verification that:

(1) π_f of a sequence is a single point.

(2) π_f identifies precisely the sequences identified by π, and maps them to the inverse orbit of 1.

Then we have the commutative diagram

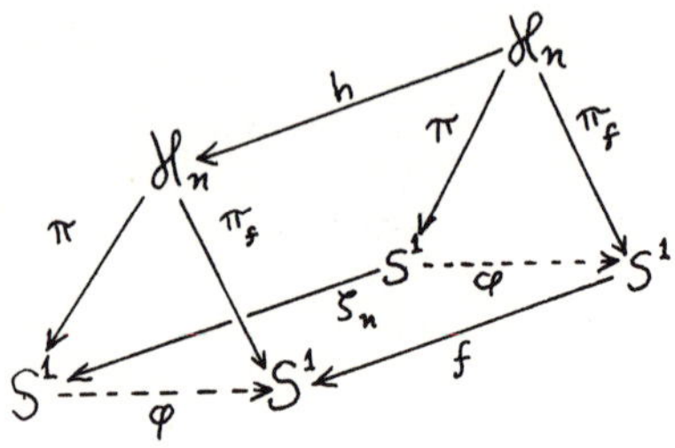

where we seek to fill in the dotted arrows by a map φ. But statement (2) above says that even where π and π_f fail to be injective, the definition

$$\varphi(p) = \pi_f(\pi^{-1}[p])$$

still makes sense as a bijection conjugating ζ_n with f. //

We would now like to take a look at what can be said of smooth endomorphisms of the circle if the "expanding" condition ($|f'| > 1$) is dropped. The first larger class of endomorphisms to be treated was the non-singular ones.

Definition: A singularity of $f: S^1 \longrightarrow S^1$ is a point x at which $f'(x) = 0$. An endomorphism f is non-singular if it has no singularities. We denote the space of non-singular C^r-endomorphisms of the circle by $N^r(S^1)$.

This case was treated by the author (Nitecki [1]). It should be pointed out that the proof of density for Axiom A given there is in error; however, the analysis of endomorphisms satisfying Axiom A is still valid. (Recall that $f \in C^r(S^1, S^1)$ satisfies Axiom A iff $\Omega(f)$ is the closure of per(f) and decomposes into a disjoint union of two closed, invariant subsets:

$\Omega(f) = \Omega_c \cup \Omega_e$

with $|f'| > 1$ on Ω_e with $|f'| < 1$ on Ω_c.) If f is C^1 but not C^2, we also assume that $f' = 1$ for only finitely many points. This is a generic property by transversality arguments. The analysis can be summarized in a

Theorem: If f $N^r(S^1)$ satisfies Axiom A and has degree > 1, then it is either an expanding map or satisfies:

(1) Ω_c is finite

(2) Ω_e has finitely many isolated points.

(3) the cluster points of Ω_e form a fully invariant cantor discontinuum

(4) on this cantor set, f is conjugate to a sub-semishift of finite type.

Furthermore, all Axiom A nonsingular endomorphisms of S^1 are C^1-structurally stable.

We will briefly sketch a proof of the structure theory using semi-conjugacies. The stability requires more work.

Our basic tool in this analysis is a slight reformulation of Shub's stability theorem, in the form on p. 65 (lemma 5).

<u>Lemma</u>: If $f \in N^r(S^1)$ is of degree $n > 1$, then there exists a monotone map h of degree 1 which commutes the diagram

$$\begin{array}{ccc} S^1 & \xrightarrow{\;f\;} & S^1 \\ {\scriptstyle h}\downarrow & & \downarrow{\scriptstyle h} \\ S^1 & \xrightarrow[\;\zeta_n\;]{} & S^1 \end{array} .$$

Here, ζ_n denotes the canonical expanding map of degree n, as before; monotonicity of h means simply that its lift to $\mathbb{R}$ is monotone in the loose sense $(x \geq y \Longrightarrow \tilde{h}(x) \geq \tilde{h}(y))$. To establish the lemma in this form, one need only check that the operator T in the proof of lemma 5 (p. 66) preserves monotonicity, and that the limit of monotone maps is again monotone.

The fact that h is monotone means that the inverse image of a point is either a single point or a closed interval. Note that if h is injective then it is automatically a homeomorphism and thus by Axiom A f is expanding; hence we will assume from the outset that h collapses some interval.

<u>Proof of Theorem</u>

We make a series of observations:

1) If $\zeta_n^k(x) = x$, then $h^{-1}[x]$ is at most a closed interval, I, with the property that $f^k[I] = I$. Thus, I must contain a fixedpoint of f^k — in fact, by invariance of domain, both endpoints of I must be fixed under f^k. Noticing that $f^k|I$ is a diffeomorphism, we can conclude that

$\Omega(f) \cap I = \mathrm{per}(f|I)$.

2) If $x \notin O^+_{\zeta_n}(x)$, that is, if x is not periodic, then the interior of $h^{-1}[x]$ (if it has any) must wander under f, since the interior is an open set with disjoint orbit. Thus,

$\Omega(f) \cap h^{-1}[x] \subset \partial h^{-1}[x]$. $(x \notin \mathrm{per}(\zeta_n))$

3) Since periodic points of ζ_n are dense, and since any neighborhood of $\partial h^{-1}[x]$ maps by h onto a set with interior, the endpoints of $\partial h^{-1}[x]$ are (by (1)) limits of periodic points for f. Thus, in case (2), we actually have

$\Omega(f) \cap h^{-1}[x] = \partial h^{-1}[x]$. $(x \notin \mathrm{per}(\zeta_n))$

4) By (1) and (3), we see that the set

$$K = \bigcup_{x \in S^1} \partial h^{-1}[x]$$

is a closed, fully invariant subset of $\Omega(f)$, and the only points of $\Omega(f)$ not in K are isolated periodic points.

5) If $x \in \mathrm{per}(\zeta_n)$ and $h^{-1}[x] = I$ has interior, then the assumption of Axiom A forces f to have finitely many periodic points in I, all of the same period. Some easy analysis shows that sinks and sources must alternate inside I, and that both endpoints of I must be sources.

6) The observation in (3) indicates that the periodic sources of f are dense in K, thus

$K \subset \Omega_e$.

7) Any x which has infinite forward orbit (ie, is not eventually periodic) under ζ_n must have $h^{-1}[x]$ a single point. (h is onto since it has degree 1). This is true because $f' > 1$ on all boundary points in the forward orbit of $h^{-1}[x]$, but if the orbit is disjoint, the lengths of the images must approach 0, so that $f' < 1$ somewhere in the forward orbit, unless the length is 0 to begin with.

8) $W^s_{loc}(\Omega_c)$ is an invariant contracting neighborhood of Ω_c; therefore by an application of the contraction lemma, Ω_c is finite. (This also follows from (5) and Axiom A.)

9) Since Ω_c is finite and intersects each open interval in $h^{-1}[\mathrm{per}(\zeta_n)]$, we see that at most finitely many $x \in \mathrm{per}(\zeta_n)$ have intervals as inverse images (under h). Caution: There may, however, be many eventually periodic points with intervals for inverse images.

10) Since isolated points of $\Omega(f)$ all lie in $S^1 \sim K$, we see that $\Omega(f)$ has finitely many isolated points.

11) If the periodic points of ζ_n with intervals for inverse images are denoted $\{x_1, \ldots, x_\ell\}$, then each interval

$I_j = \text{interior } h^{-1}[x_j]$

consists of a "block" with finitely many periodic points. A maximal interval in $S^1 \sim K$ eventually maps (under f) onto one of the I_j ; if it is not itself one of the I_j, then it consists of wandering points.

12) To set up the conjugacy of $f|K$ with a sub-semishift, one sets up a partition as in the final proof of Shub's stability theorem (one can, in fact, simply take one point p_m in $h^{-1}[\frac{m}{n}]$ for each $m = 0, \ldots, n-1$), adding a few more partition points, however, so that the endpoints of the intervals $\{h^{-1}[x_1], \ldots, h^{-1}[x_\ell]\}$ as in (11) are separated by the partition. One then sets up symbolism as before, this time employing more than the n original symbols (since there are now more than n intervals in the partition). An "admissible" subset of finite type among one-sided symbol sequences can then be defined by eliminating any pair $\alpha\beta$ such that for the corresponding intervals, $f\mathring{A}_\alpha \cap \mathring{A}_\beta = \emptyset$. With a little care, it is then shown that the projection π_f can be defined on admissible sequences by the formula

$$\pi_f\{a_i\} = \bigcap_{i \geq 0} f^{-i}[A_i] .$$

As before, one shows that the image of a sequence is a single point, and the addition of partition points to separate endpoints of $h^{-1}[x_j]$ insures in fact that the map π_f is one-to-one. //

The stability follows from the observation that the partition we have set up works as well for perturbations of the map, but as we said before, the details of the partition argument require considerable work.

APPENDIX II

THE KUPKA-SMALE THEOREM

We prove here the rest of the Kupka-Smale theorem which we began in theorem 4, 2.2. The full statement of the theorem is:

Theorem (Kupka-Smale): For a residual subset of $\text{Diff}^r(M)$, the following properties hold:
(1) all periodic points are hyperbolic
(2) for any periodic point p, $W^s(p)$ and $W^u(p)$ are injectively immersed hyperplanes
(3) for any two periodic points, p and q, $W^s(p) \pitchfork W^u(q)$.

Historically, this theorem was proven independently by Kupka [1] in a flow formulation and Smale [7] in both forms, after Markus [1] had found partial results; other proofs, formulated in terms of flows, are given by Abraham-Robbin [1] and Peixoto [2], who gives a generalization to M non-compact. Finally, an endomorphism version of the theorem is given by Shub [1].

The proof we give is based on muddled readings of some of these references and conversations with M. Shub and J. Palis.

We recall that theorem 4 in 2.2 is exactly statement (1) of the theorem. Statement (2) is a consequence:

Lemma 1: If $f \in \text{Diff}^r(M)$ and x is a hyperbolic fixedpoint, then $W^u(x)$ is an injectively immersed hyperplane.

Proof

Recall that the unstable manifold theorem (theorem 2, 5.2) gives us the local unstable manifold $W^u_{loc}(x)$ as an embedded smooth disc, and by definition

$$W^u(x) = \bigcup_{n \geq 0} f^n W^u_{loc}(x).$$

It is clear that $W^u(x)$ is immersed; we need to establish that the immersion has no self-intersections.

Suppose there were a self-intersection; since $W^u_{loc}(x)$ is embedded, this would mean that for some iterate of f and two points $y \neq z \in W^u_{loc}(x)$, $f^m y = f^m z$. But since f is a diffeomorphism, this is impossible. //

Of course, the corresponding lemma holds for $W^s(x)$ (consider f^{-1}).

To prove the genericity of statement (3), we take an approach similar to that used in handling hyperbolicity. We show that the corresponding statement, restricted to periodic points of bounded period, holds for a dense open set.

To this end, we let

$$D_n = \{f \in \bigcap_{i=1}^{n} H_i \mid \forall p,q \in \text{per}(f) \text{ with period } \leq n, \\ W^s(p) \pitchfork W^u(q)\}.$$

Openness is the easier result.

<u>Lemma 2:</u> D_n is open in $\text{Diff}^r(M)$.

<u>Proof</u>

Suppose $f \in D_n$; we need to show any C^1-perturbation g is also in D_n. Let p, q be periodic for g with period $\leq n$. By openness of H_n, p and q are both hyperbolic so that $W^s(p)$, $W^u(q)$ are immersed planes. Since $W^u(q)$ is invariant and g is a diffeomorphism, it will suffice to show that $W^s_{loc}(p) \pitchfork W^u(q)$ - in fact, it suffices to show that for any closed disc E about p in $W^s_{loc}(p)$,

$E \pitchfork W^u(q)$.

However, since E can be taken compact, this last statement easily follows from the fact that (stable and unstable) manifolds vary continuously with f. Since there are a finite number of periodic points for f or g with given period, this completes the proof. //

Lemma 3: D_n is dense in $\mathrm{Diff}^r(M)$.

Proof

By previous analysis, we can assume $f \in \bigcap_{i=1}^{n} H_n$ and immediately reduce the problem to showing that if $p, q \in \mathrm{per}(f)$ with period $\leq n$, then for some nearby map g, $p, q \in \mathrm{per}(g)$ and $W^s(p) \pitchfork W^u(q)$. It will suffice to show that for any m, the set

$$D_{n,m} = \{f \in \bigcap_1^n H_i \mid (\forall p,q \text{ with period } \leq n) \;\; W^s_{loc}(p) \pitchfork f^i W^u_{loc}(q) \;\; \forall i \leq m\}$$

is dense in H_n, since $D_n = \bigcap_{m=1}^{\infty} D_{n,m}$.

If $f \in \bigcap_1^n H_i$ and $p, q \in \mathrm{per}(f)$, periods $\leq n$, with $W^s_{loc}(p) \pitchfork f^i W^u_{loc}(q)$ ($\forall\, i \leq m-1$) but $W^s_{loc}(p)$ intersects $f^m W^u_{loc}(q)$ non-transversally at fx, we note that $x \notin f^{m-2} W^u_{loc}(q)$, which means we can pick a neighborhood U of x in M which intersects neither W^s_{loc} nor $f^{m-2} W^u_{loc}(q)$ and modify f only on U (using transversality theory) so that

$f|[U \cap f^{m-1}W^u_{loc}(q)] \pitchfork W^s_{loc}(p)$. Note that we haven't changed f near $f^{m-2}W^u_{loc}(q) \cup W^s_{loc}(p)$, so that such a process of perturbation applied to each "bad" pair p, q for $f \in D_{n,m}$ in turn will finally lead to a perturbation $g \in D_{n,m+1}$. //

The Kupka-Smale Theorem now follows from noting that diffeomorphisms satisfying (1), (2) and (3) include all elements of the residual set $D_\infty = \bigcap_{n=1}^{\infty} D_n = \bigcap_{n,m=1}^{\infty} D_{n,m}$.

APPENDIX III

PERIODIC POINTS OF TORAL AUTOMORPHISMS

In this appendix we give an alternate proof of lemma 1 in 3.1. This proof, given by Banchoff and Rosen [1], again uses the automorphism properties of a map but it also exhibits the periodic points explicitly.

Definition: A toral automorphism is a diffeomorphism of $T^n = \mathbb{R}^n/\mathbb{Z}^n$ induced by a linear automorphism of $\mathbb{R}^n$ of determinant ± 1 which preserves the integer lattice $\mathbb{Z}^n$.

The result we prove is then

Lemma: For any toral automorphism f, all points with rational coefficients are periodic. If f is hyperbolic, then these give all the periodic points.

Proof

We work with matrices. f is induced by a matrix $f_o \in GL(n,\mathbb{Z})$. The condition that $x \in \mathbb{R}^n$ project to a periodic point of period m is equivalent to

$$f_o^m x - x = (f_o^m - I)\, x \in \mathbb{Z}^n . \qquad (1)$$

If f_o is hyperbolic and x satisfies (1), then the observation that $f_o^m - I$ is a non-singular integer matrix together with Cramer's rule give that $(f_o^m - I)^{-1}$ is a matrix with rational entries, so that

$$x = (f_o^m - I)^{-1}[f_o^m x - x] \in (f_o^m - I)^{-1}\, \mathbb{Z}^n \subset \mathbb{Q}^n$$

has rational coefficients.

Conversely , if $x \in \mathbb{Q}^n$, we let r be the lowest common denominator of the coefficients of x, and consider the matrix $\phi_r(f_o)$ obtained from f_o by reducing all entries mod r. Then $\phi_r(f_o)$ is a matrix with entries in the ring $\mathbb{Z}_r$ of integers mod r, and in fact ϕ_r induces a ring homomorphism $\phi_r : GL(n,\mathbb{Z}) \longrightarrow GL(n,\mathbb{Z}_r)$.

But $GL(n,\mathbb{Z}_r)$ is a finite group, so

$$\phi_r(f_o^m) = \phi_r(f_o)^m = I$$

for some m, and hence

$$\phi_r(f_o^m = I) = 0$$

or

$$f_o^m - I = 0 \mod r$$

which means all entries are integer multiples of r, so that

$$f_o^m x - x \in \mathbb{Z}^n ,$$

proving the lemma. //

In [1], Banchoff and Rosen extend this result to infra-nilmanifold automorphisms.

APPENDIX IV

OPENNESS OF ANOSOV DIFFEOMORPHISMS

We sketch here an approach to showing that Anosov diffeomorphisms are an open subset of $\mathrm{Diff}^1(M)$.

First, a word of caution: it is very easy to mistakenly deduce this fact as a corollary of Lemma 3 in 3.2. One reasons: f is Anosov iff f^* is hyperbolic; if g is C^1-near f, then g^* is near f^*, and so, by the openness of linear hyperbolic maps, g^* is hyperbolic, hence g is Anosov. The problem is, that when we deal with spaces of non-Lipschitz maps, any two pre-composition operators, α_φ and α_ψ, are at distance 1, so that g^* need not be near f^*, no matter how close g is to f in the C^r sense. (Thanks to P. Walters for showing me this error.) Furthermore, the idea of trying to pass to the space Γ^r instead of Γ_b as a domain for f^* is likely to fail, because otherwise it would probably follow that the Whitney sum $E^s \oplus E^u$ is smooth, a fact known to be false in general (see e.g. Anosov).

Thus, a different approach is called for. We go back to a slightly modernized version of Anosov's original approach to the problem - the techniques of stable manifold theory.

We are given f an Anosov diffeomorphism of M, with

splitting $TM = E^s \oplus E^u$, and wish to show that any sufficiently C^1-nearby map g will have an invariant, hyperbolic splitting $TM = \tilde{E}^s \oplus \tilde{E}^u$. We shall find $\tilde{E}^u$; the stable bundle can be found by an adaptation of the same techniques.

As in the case of stable manifold theory, we express the line bundle $\tilde{E}^u$ as the graph of a linear map $\varphi_g : E^u \longrightarrow E^s$ in each fiber. More formally, let $\mathcal{L}$ denote the unit ball bundle in the vector bundle whose fiber at each point $x \in M$ is the space $L(E^u_x, E^s_x)$ of linear maps from the unstable fiber (of f) to the stable fiber. We will define, for g near f, a "graph transform" Γ_g acting on sections of $\mathcal{L}$ and show that Γ_g has a fixedpoint.

A technical difficulty arises from the fact that in this case we are considering linear "graphs" which lie, not in a fixed Banach space, but in varying fibers of the tangent bundle. However, we can still build up a "graph transform" on sections of $\mathcal{L}$, as follows.

First, we note that the image of a section of a bundle is equivalent to its graph: the fact that $\pi\gamma(x) = x$ insures that two sections with the same image are actually the same map. Thus, a section σ of $\mathcal{L}$ is totally determined by a subset S of the total space $\mathcal{L}$ with the property that S contains a unique point in each fiber of $\mathcal{L}$. Furthermore, a point of $\mathcal{L}$ is simply a hyperplane (of slope ≤ 1) in some fiber of $TM = E^s \oplus E^u$. Tf and Tg each act on such hyperplanes; Tf takes a hyperplane in T_xM to one in $T_{fx}M$, and leaves the section E^u of $\mathcal{L}$ fixed. Tg, on the other hand, takes a hyperplane in T_xM to one in $T_{gx}M$. To in some way compare the action of Tg to that of Tf, we express $Tg_x\colon T_xM \longrightarrow T_{gx}M$ by a matrix of operators

$$Tg_x = \begin{pmatrix} A_x & B_x \\ C_x & K_x \end{pmatrix}$$

where

$$\begin{aligned} A_x&\colon E^u_x \longrightarrow E^u_{gx} \\ B_x&\colon E^s_x \longrightarrow E^u_{gx} \\ C_x&\colon E^u_x \longrightarrow E^s_{gx} \\ K_x&\colon E^s_x \longrightarrow E^s_{gx}\,. \end{aligned}$$

Then when g is C^1-near f, $\|C_x\|$ and $\|B_x\|$ are nearly 0; $\|K_x\|$ is near τ, the skewness of Tf, and $\|A_x\|$ has minimum norm near τ^{-1}. Thus, Tg is a perturbation of the linear operator L_g given by the matrix

$$\begin{pmatrix} A_x & 0 \\ 0 & K_x \end{pmatrix}$$

which is "hyperbolic" on sections of $\mathcal{L}$. We can mimic the arguments of graph transform theory, using L_g in place of the linear map T, to deduce the existence of a fixed bundle $\tilde{E}^u$ for Tg.

The above sketch is, of course, only a plan of action; one additional observation will probably make it clear that the "mimicking" can in fact be accomplished.

On the fibers of TM, the map L_g preserves the original splitting $E^s \oplus E^u$, contracting E^s and expanding E^u. It thus defines a "graph transform" $\tilde{\Gamma}_{g,x}$ from one fiber of $\mathcal{L}$ to another, characterized by:

$$\varphi \in \mathcal{L}_x, \text{ ie, } \varphi : E^u_x \longrightarrow E^s_x, \text{ then } \tilde{\Gamma}_{g,x}(\varphi) : E^u_{gx} \longrightarrow E^s_{gx},$$
$$\text{with } gr(\tilde{\Gamma}_{g,x} \quad) = L_g\, gr(\varphi) .$$

This map of points of $\mathcal{L}$ to points of $\mathcal{L}$ yields a linear set-mapping on images of sections, which in turn induces a linear map $\mathcal{L}_g$ on the sections of $\mathcal{L}$; if γ is a section of $\mathcal{L}$ (ie, for each $x \in M$, γ_x is a linear map $E^u_x \longrightarrow E^s_x$ of norm ≤ 1) then the image of γ (as a subset of $\mathcal{L}$) is taken to the image of the section

$$\mathcal{L}_g(\gamma)_x = \tilde{\Gamma}_{g,x}(\gamma_{g^{-1}x})$$

This operator on sections of $\mathcal{L}$ is hyperbolic; the operator induced by Tg instead of L_g is then a Lipschitz perturbation of this, and hence has an invariant section, $\tilde{E}^u$.

The details of this argument simply consist of a translation of the estimates and arguments of section 5.2 to this case. (After constructing $\tilde{E}^u$, one constructs $\tilde{E}^s$ as the "unstable" bundle for g^{-1}.)

We obtain:

<u>Proposition</u>: The C^1 Anosov diffeomorphisms of M are an open subset of $\text{Diff}^1(M)$.

BIBLIOGRAPHY

ABBREVIATIONS

1 Global Analysis = Global Analysis. Proc. Symp. Pure Math., vol. 14, AMS, 1970 (Proceedings of the Summer Institute on Global Analysis, Berkeley, Calif., 1968).

2 BAMS = Bulletin of Amer. Math. Soc.

3 TAMS = Transactions of Amer. Math. Soc.

4 PAMS = Proceedings of Amer. Math. Soc.

Abraham, R.

1 Lectures of Smale on Differential Topology, Mimeographed Notes, Columbia, 1962

2 "Transversality in manifolds on mappings," BAMS, vol. 69 (1963), pp. 470-474

______, and Marsden, J.

1 Foundations of Mechanics. NY: Benjamin, 1967.

______, and Robbin, J.

1 Transversal Mappings and Flows. NY: Benjamin, 1967.

______, and Smale S.

1 "Non-genericity of Ω-stability," Global Analysis, pp. 5-8.

Alekseev, V. M.

1 "Quasirandom dynamical systems I. Quasirandom diffeomorphisms," Math. USSR Sbornik vol. 5 (1968), pp. 73-128.

Andronov, A. and Pontryagin, L.

1 "Systemes grossiers," Dokl. Akad. Nauk. SSSR, vol. 14 (1937), pp. 247-251.

Anosov, D. V.

1 "Roughness of geodesic flows on compact Riemannian manifolds of negative curvature," Soviet Math. Dokl. vol. 3 (1962), pp. 1068-1070.

2 "Ergodic properties of geodesic flows on closed Riemannian manifolds of negative curvature" Soviet Math. Dokl. vol. 4 (1963), pp. 1153-1156.

3 "Geodesic flows on compact Riemannian manifolds of negative curvature," Proc. Steklov Math. Inst. vol. 90 (1967).

______, and Sinai, Ja. G.

1 "Certain smooth ergodic systems," Russian Math. Surveys, vol. 22 (1967), pp. 103-167.

Arnold, V. I., and Avez, A.

1 Ergodic Problems of Classical Mechanics NY: Benjamin, 1968

______, and Sinai, Ja. G.

1 "Small perturbations of the autmorphisms of the torus," Soviet Math. Dokl., vol. 3 (1962) pp. 783-787 correction, vol. 4 (1963) p. vi.

Artin, E., and Mazur, B.

1 "On periodic points," Annals of Math., vol. 81 (1965), pp. 82-99.

Auslander, L.

1 "Bieberbach's Theorem on Space Groups and Discrete Uniform Subgroups of Lie Groups," Annals of Math., vol. 71 (1960) pp. 579-590.

Banchoff, T., and Rosen, M.

1 "Periodic points of Anosov diffeomorphisms," Global Analysis, pp. 17-21.

Baxendale, P.

1 "Classification of Markov partitions," mimeo notes, University of Warwick, Coventry, 1970.

Bonic, R., and Frampton, E.

1 "Smooth Functions on Banach Manifolds," J. Math. and Mech., vol. 15 (1966), pp. 877-898.

Bowen, R.

1 "Topological entropy and Axiom A," Global Analysis, pp. 23-42.

2 "Periodic points, Measures and Axiom A," Mimeographed Notes, University of Warwick, Coventry, England (to appear, TAMS)

3 "Markov partitions for Axiom A diffeomorphisms," Am. J. Math. vol 92 (1970) pp. 725-747.

4 "Markov partitions and minimal sets for Axiom A diffeomorphisms," Am. J. Math., vol. 92 (1970), pp. 907-918.

Byers, W.

1 Thesis, University of California, Berkeley, 1969

Coddington, E. A., and Levinson, N.

1 Theory of Ordinary Differential Equations, NY (McGraw-Hill) 1955.

Denjoy, A.

1 "Sur les courbes définies par les équations différentielles á la surface do tore." J. de Math. Pures et Appliquées v. 11

(ser 9) (1932) pp. 333-375.

2 "Les trajectoires á la surface du tore" Comptes Rendus Acad. Sci. vol. 223 (1946) pp. 5-8

3 Theorie des fonctions sur les characteristiques a la surface du tore, Comptes Rendus Acad. Sci. vol. 194 (1932) pp. 830-833.

4 Theorie des fonctions sur les characteristiques du tore, Comptes Rendus Acad. Sci. vol. 194 (1932) pp. 2014-2016.

Eells, J.

1 "On the Geometry of Function Spaces," in Symposium Internacional de Topologia Algebraica Mexico, 1956.

2 "A Setting for Global Analysis" in BAMS, v 72 (Sept. 1966) pp. 751-807.

Fenichel, N.

1 "Persistence and Smoothness of invariant manifolds for flows," to appear, Indiana University Mathematics Journal.

Field, M.

1 "G-invariant dynamical systems," mimeo notes, University of Warwick, Coventry, 1970.

Franks, J.

1 Anosov Diffeomorphisms, Global Analysis, pp. 61-94.

2 "Anosov Diffeomorphisms on Tori," TAMS, vol. 145 (1969) pp. 117-124.

3 "Necessary Conditions for Stability of Diffeomorphisms," to appear.

4 "Differentiably Ω-stable diffeomorphisms," to appear.

5 "C^K-stability of mappings," to appear.

Gottschalk, W., and Hedlund, G. A.

1 Topological Dynamics, AMS Colloquium publications, vol. 36, Providence, R. O., 1955.

Grobman, D.

1 "Homeomorphisms of systems of differential equations," Dokl. Akad. Nauk. vol. 128 (1965), pp. 880-881.

Guckenheimer, J.

1 "Endomorphisms of the Riemann Sphere," <u>Global Analysis</u>, pp. 95-124.

2 "Axiom A + no cycles $\Longrightarrow$ $\zeta_f(t)$ rational," BAMS, vol. 76 (1970), pp. 592-5.

Hartman, P.

1 Ordinary Differential Equations, NY: Wiley, 1964.

Hirsch, M. W.

1 "Expanding maps and transformation groups," <u>Global Analysis</u> pp. 125-132.

2 "On invariant subsets of hyperbolic sets," <u>Essays on Topology and Related Topics</u>, Berlin: Springer-Verlag (1970), pp. 126-135.

3 "Foliations and noncompact transformation groups," BAMS , vol. 76 (1970), pp. 1020-1023.

______, Palis, J., Pugh, C., and Shub, M.

1 "Neighborhoods of Hyperbolic Sets," Inventiones Math. vol. 9 (1970), pp. 121-134.

______, and Pugh, C.

1 "Stable manifolds and hyperbolic sets," <u>Global Analysis</u>, pp. 133-165.

Hirsch, M. W., Pugh, C., and Shub, M.

1 Invariant Manifolds, To appear. Announcement: BAMS, vol. 76 (1970) pp. 1015-1019.

Kopell, N.

1 "Commuting diffeomorphisms," Global Analysis, pp. 165-184.

Krzyżewski, K., and Szlenk, W.

1 "On invariant measure for expanding differentiable mappings," Studia Math., vol. 33 (1969), pp. 83-92

Kupka, I.

1 "Contribution à la théorie des champs génériques," Contributions to Differential Equations, vol. 2 (1963), pp. 457-484 and vol. 3 (1964), pp. 411-420.

Manning, A.

1 "Axiom A diffeomorphisms have rational zeta functions," mimeographed notes, University of Warwick, Coventry, 1970.

Markley, N.

1 "Homeomorphisms of the circle without periodic points," Proc. London Math. Soc., 3d. series, vol. 20 pp, 688-698.

Mather, J.

1 "Characterization of Anosov diffeomorphisms," Indag. Math. vol. 30 (1968), pp. 479-483.

Markus, L.

1 "Structurally stable differential systems," Ann. of Math. (2) vol. 73 (1961), pp. 1-19.

Meyer, K.

1 "Periodic Points of diffeomorphisms," BAMS, vol. 73 (1967) pp. 615-17.

2 "On the convergence of the zeta function for flows and diffeomorphisms," J. Diff. Eq., vol. 5 (1969), pp.338-345.

Morse, M.

1 Symbolic Dynamics, Mimeographed Notes by R. Oldenberger, Institute for Advanced Study, Princeton: 1966.

Moser, J.

1 "On a Theorem of D. Anosov," J. Diff. Eq., vol. 5 (1969) pp. 411-440.

Munkres, J. R.

1 Elementary Differential Topology. Annals of Mathematics Studies, No. 54, Princeton University Press, 1966.

Nemytskii, V. V., and Stepanov, V. V.

1 Qualitative Theory of Differential Equations. Princeton, Princeton University Press, 1960.

Newhouse, S.

1 "On codimension one Anosov diffeomorphisms," Am. J. Math., vol. 92 (1970), pp. 761-770.

2 "Non-density of Axiom A(a) on S^2", Global Analysis, pp. 191-202.

3 "Hyperbolic limit sets," to appear.

Nitecki, Z.

1 "Non-singular endomorphisms of the circle," Global Analysis, pp. 203-220.

Palais, R. S.

1 Lectures on Differential Topology. Mimeographed notes, Brandeis University, 1964-5.

2 Lectures on Morse Theory. Mimeographed Notes, Brandeis University.

3 Foundations of Global Non-Linear Analysis. NY: Benjamin, 1968

4 "Extending diffeomorphisms," PAMS, vol. 11 (1960), pp. 274-77.

Palis, J.

1 "On the local structure of hyperbolic points in Banach spaces." Anais da Academia Brasileira de Ciêncas (1969)

2 "On Morse-Smale Dynamical Systems" Topology 4 (1969) pp. 385-404.

3 "A note on Ω-stability" Global Analysis, pp. 221-2.

______, and Smale, S.

1 "Structural Stability Theorems," Global Analysis, pp. 223-232.

Peixoto, M. M.

1 "Structural Stability on two-dimensional manifolds," Topology vol. 1 (1962), pp. 101-120

2 "On an approximation theorem of Kupka and Smale," J. Diff. Eq., vol. 3 (1966) , pp. 214-227.

______, and Pugh, C.

1 "Structurally Stable systems on open manifolds are never dense", Annals of Math., vol 87 (1968), pp. 423-30.

Plante, J.

1 "Anosov flows," to appear.

2 "Diffeomorphisms with invariant line bundles," to appear.

Poincaré, H.

1 New Methods of Celestial Mechanics. (Transl. by Anon. of "Les Méthodes Nouvelles de la Mécanique Céleste," Paris: 1892-99), NASA, 1967.

Pugh, C.

1 "The Closing Lemma," Am. J. Math., vol. 89 (1967), pp. 956-1009.

2 "An Improved Closing Lemma and a General Density Theorem," Am. J. Math., vol. 89 (1967), pp. 1010-1021.

3 "On a theorem of P. Hartman," Am. J. Math., vol. 91 (1969), p. 363.

______, and Robinson, R. C.

1 "The C^1 closing lemma," to appear.

______, and Shub, M.

1 "Ergodicity of Anosov actions," to appear.

Riesz, F., and Sz.-Nagy, B.

1 Functional Analysis. NY: Ungar, 1955.

Robbin, J.

1 "A Structural Stability Theorem," to appear, Annals of Math. Preliminary announcement: "On structural stability," to appear, BAMS.

Robinson, R. C.

1 "Generic properties of conservative systems," Am. J. Math., vo . 92 (1970), pp. 562-603.

2 "Generic properties of conservative systems, II," Am. J. Math., vol. 92 (1970), pp. 897-906.

Sacker, R.

1 "On periodic solutions near homoclinic points - a new treatment of Smale's work" Mimeographed lecture notes from a Seminar on Ordinary Differential Equations at NYU (Courant Institute?) April, 1965.

2 "A perturbation theorem for invariant Riemannian manifolds," in Hale and LaSalle (ed.), Differential Equations and Dynamical Systems, Proc. of the Int. Symp., Mayaguez, Puerto Rico, 1965; NY: Academic Press, 1967, pp. 43-54.

Schneider, C.

1 Thesis, University of California, Berkeley, 1969.

Shahshahani, S.

1 Thesis, University of California, Berkeley, 1969.

Shub, M.

1 "Endomorphisms of Compact Differentiable Manifolds," Am. J. Math., vol. 91 (1969), pp. 175-199.

2 "Periodic orbits of hyperbolic diffeomorphisms and flows," BAMS, vol. 75 (1969), pp. 57-8.

______, and Williams, R.

1 "Future stability is not generic," Proc. AMS, vol. 22 (1969), pp. 483-4.

Siegel, C. L.

1 "Note on differential equations on the torus," Annals of Math., vol. 46 (1945), pp. 423-428.

Simon, C.,

1 "Non-genericity of rational zeta functions and instability in $\mathrm{Diff}^r(T^3)$," Thesis, Northwestern University, 1970.

Sinai, Ja. G.

1 "Markov Partitions and C-diffeomorphisms," Functional Analysis and its Applications, vol. 2 (1968), pp. 61-82.

2 "Theory of Dynamical Systems, part 2," to appear as Lecture Notes, Aarhus Universitet, Denmark.

Smale, S.

1 "Morse inequalities for a dynamical system," BAMS vol. 66 (1960), pp. 43-49.

2 "On gradient dynamical systems," Annals of Math., vol. 74 (1961), pp. 199-206.

3 "Differentiable Dynamical Systems," BAMS, vol. 73 (1967) pp. 747-817.

4 "Diffeomorphisms with many periodic points," in Differential and Combinatorial Topology (S, S, Cairns, ed.) Princeton University Press, 1965. pp. 63-80.

5 "The Ω-Stability Theorem," Global Analysis, pp. 289-298.

6 "Structurally stable systems are not dense," Am. J. Math., vol. 88 (1966), pp. 491-496.

7 "Stable manifolds for differential equations and diffeomorphisms," Ann. Scuola Normale Superiore Pisa, vol. 18 (1963), pp. 97-116.

Sondow, J.

1 "Fixed points of Anosov maps of certain manifolds," to appear.

Thom, R.

1 "Un Lemma sur les applications différentiables," Bol. Soc. Mat. Mex. (2), vol. 1 (1956), 59-71.

Tomter, P.

1 "Anosov flows on infrahomogeneous spaces," Global Analysis, pp. 299-328.

Van Kampen, E. R.

1 "The topological transformations of a simple closed curve into itself," Am. J. Math., vol. 57 (1935), pp. 142-152.

Walters, P.

1 "Anosov diffeomorphisms are Topologically Stable," Topology vol. 9 (1970), pp. 71-78.

Williams, R. F.

1 "The 'DA' maps of Smale and structural stability," Global Analysis, pp. 329-334.

2 "One dimensional non-wandering sets," Topology, vol. 6 (1967), pp. 473-478.

3 "Classification of one dimensional attractors," Global Analysis, pp. 341-362.

4 "Expanding attractors," Mimeographed notes, University of Warwick, Coventry, England.

5 "Classification of Expanding Attractors," to appear.

6 "Classification of Symbol Spaces of Finite Type," to appear.

INDEX